RAUL CANALES ECOBEDO
JOSÉ LUIS RIOS FLORES
MANUEL DE JESUS A. RUIZ-ESPARZA

PHYSICAL, ECONOMIC AND SOCIAL WATER FOOTPRINT

RAUL CANALES ECOBEDO
JOSÉ LUIS RIOS FLORES
MANUEL DE JESUS A. RUIZ-ESPARZA

PHYSICAL, ECONOMIC AND SOCIAL WATER FOOTPRINT

OF APPLE(Malus domestica L.) RED DELICIOUS VARIETY FROM NORTHERN MEXICO

ScienciaScripts

Imprint

Any brand names and product names mentioned in this book are subject to trademark, brand or patent protection and are trademarks or registered trademarks of their respective holders. The use of brand names, product names, common names, trade names, product descriptions etc. even without a particular marking in this work is in no way to be construed to mean that such names may be regarded as unrestricted in respect of trademark and brand protection legislation and could thus be used by anyone.

Cover image: www.ingimage.com

This book is a translation from the original published under ISBN 978-613-9-40476-6.

Publisher:
Sciencia Scripts
is a trademark of
Dodo Books Indian Ocean Ltd. and OmniScriptum S.R.L publishing group

120 High Road, East Finchley, London, N2 9ED, United Kingdom
Str. Armeneasca 28/1, office 1, Chisinau MD-2012, Republic of Moldova, Europe
Managing Directors: Ieva Konstantinova, Victoria Ursu
info@omniscriptum.com

Printed at: see last page
ISBN: 978-620-8-08599-5

TABLE OF CONTENTS

SUMMARY

The objective was to determine the water footprint (HHF) in its three variants: physical, economic and social through physical (EFA), economic (EEA) and social (ESA) water efficiency indexes, as well as the physical (PFA), economic (PEA) and social (PSA) productivity of water used in the production of the apple crop (*Malus domestica L.)* Red Delicious variety produced in northern Mexico, in the main producing states: Chihuahua, Coahuila and Durango in 2022. Mathematical models of water efficiency and productivity were used for individual crops, as well as at the aggregate level for several crops. The independent variables feeding the models were commercial-scale production data. Always in the order of Durango, Coahuila, Chihuahua and the weighted average of the three, the results indicate: in EFA (in liters/kg):6346, 1111, 391 and 541; in FSS (in m^3 /USD gain (+) or loss (-): -1.82, -0.56, 3.27 and -99.63: in ESA (in m^3 /employment): 65829, 13091, 18680 and 22570; in ADP (in kg/m^3): 0.158, 0.900, 2.557 and 1.848; in EAP (in kg/m): 0.158, 0.900, 2.557 and 1.848; in EAP

(in USD gain(+) or loss (-) /m^3): -0.55, +1.78, 0.31 and -0.01; in PSA (in jobs/hm^3):15.19, 76.39, 53.33 and 44.31. Only the physical WF of Chihuahua and the weighted average of the three states (with 389 and 541 L kg^{-1} respectively) were below the world average WF of 700 L kg^{-1} , Durango and Coahuila with 6346 and 1,111 were above the 700 L kg of the world average WF, which places these two states in a bad position, It also shows them as inefficient and unproductive in the use of water in physical and economic terms (as they produced a loss per hm^3 used) and only Coahuila comes out well in the social WF, as it is the one that uses the least water per job generated.

Key words: apple - productivity - efficiency - virtual water - sustainability - water footprint.

ABSTRACT

The objective was to determine the water footprint (HHF) in its three variants: physical, economic and social through indices of physical efficiency (EFA), economic (EEA) and social (ESA) of water, as well as physical productivity (PFA), economic (PEA) and social (PSA) of the water used in the production of the apple crop (*Malus domestica L.*) variety Red Delicious produced in the north of Mexico, in the main producing states: Chihuahua, Coahuila and Durango in 2022. The mathematical models of water efficiency and productivity for an individual crop as well as at an aggregate level for several crops. The independent variables that fed into the models were commercial-scale production. Always in the order of Durango, Coahuila, Chihuahua and the weighted average of the three, the results indicate: in the EFA (in liters/kg): 6346, 1111, 391 and 541; in the EEA (in m^3 /USD of profit (+) or loss (-): -1.82, -0.56, 3.27 and -99.63: in the ESA (in m^3 /employment): 65829, 13091, 18680 and 22570; in the PEA (in kg/m^3): 0.158, 0.900, 2.557 and 1.848; in the PEA (in USD gain(+) or loss (-)/m3): -0.55, +1.78, 0.31 and -0.01; in the PSA (in jobs/hm^3): 15.19, 76.39, 53.33 and 44.31. Only the Physical HH of Chihuahua and the weighted average of the three states (with 391 and 541 L kg^{-1} respectively) were located below the world average HH of 700 L kg^{-1} , Durango and Coahuila with 6346 and 1,111 were above the 700 L kg^{-1} of the world average WF, which poorly positions these two states, since it suggests an enormous transfer of Mexican water to other countries crystallized in the form of the Red Delicious apple, as well as to the domestic market itself, in turn points them out as inefficient and unproductive in the use of water in physical and economic terms (since they produced loss per hm^3 used) and only Coahuila It comes out well in the social HH, since it is the one that uses the least water per employment generated.

Keywords: apple -productivity -efficiency - virtual water - sustainability

- Water footprint.

I.   INTRODUCTION

The apple, undoubtedly one of the most popular fruits in the world, not only for its exquisite flavor, but also for its remarkable health benefits, as shown by BUPA, 2023[1] . It has the following benefits: "Strengthens the immune system

Improves brain function

Helps to solve intestinal problems and stomach pain Prevents diabetes (or helps to control it)

Protects against cardiovascular diseases. Fights against asthma

Prevents cavities.

Helps to prevent cancer, heart disease, anemia Improve mood

Prevent early aging" Precisely because of its great palatability and its healthy properties, apples are the main so-called superfoods, which has caused, according to 2000 Agro Agro Industrial Magazine, 2021[2] . It is known that globally apple production increased at an average annual rate of 8.4 percent in the period 2012-2020, while in Mexico in that period the production of this fruit had an annual growth rate of 14%.

Now, as the demand for apples is growing, implicitly more resources will have to be used in its production, specifically it will imply a greater demand for fresh water in its production, given the problem to be addressed where this work is inserted, that of the scarcity of water, so necessarily, in a necessary and objective way, it is necessary to determine the degree of efficiency and productivity with which scarce fresh water is used in this super food called apple, It is necessary to

---

[1] BUPA (British United Provident Association), 2023. All the benefits of eating apples. Available at https://www.bupasalud.com/agentes/novedades/actualizaciones-2023
Date of last consultation January 7, 2024.
[2] 2000 Agro Agro Industrial Magazine, 2021. Available at:
https://www.2000agro.com.mx/agroindustria/crecimiento-de-2-digitos-en-produccion-de-manzanas/

determine the degree of efficiency and productivity with which the scarce fresh water is used in this super-food called apple, because although it is a healthy and tasty food, this does not exempt it from making an adequate use of such a scarce resource as water, since not only the sustainability of production in the long term depends on it: the availability of water for future generations for human consumption is at stake, which is why the proper use of water is so important.

It should be clear from the outset that this work does not deal with the cultivation of Red Delicious apples, but rather with how the main resource in agricultural production is used in the production of Red Delicious apples: fresh water, which, as will be seen later, is a really scarce resource, since the fresh water *available* for human beings barely represents 0.0075% of the total existing water on the planet, hence the importance of having index numbers that clearly indicate, in **physical terms**, how much water is needed for each kilogram of fruit produced, or what is the same but in reverse: how many kilograms are produced for each cubic meter of water used, likewise, in **economic terms**, how much profit is generated by each cubic meter of water used in production, or said as its inverse, how many cubic meters of water are needed to produce a monetary unit of profit, and finally, in **social terms**, how much employment is generated by using a cubic hectometer of water in production or seen as its inverse: how many cubic meters of water is associated with the creation of a job in the agricultural production of Red Delicious apple.

In this way, only when we have the index numbers of the water footprint in physical, economic and social terms of the whole set of crops of an agricultural region, we are able to make an adequate allocation of water among the different alternatives in which water can be used, in order to use it, always having an objective when using water, to allocate water to those productive activities in crops or livestock activities, which, while maximizing economic and social benefits, in physical terms save water, thus contributing to long-term sustainability and the availability of the resource for use by future generations.

II. SPECIFIC OBJECTIVES AND HYPOTHESES

2.1 Objectives

The general objective of this study was to determine the water footprint in its three facets: in physical terms, in economic terms and in social terms, through mathematical equations fed with commercial scale data, in order to determine, firstly, ***what is the amount of product*** (physical - in kg-, economic -in USD profit, and social -in jobs generated-) *that is produced for each cubic meter of water used in the production of Red Delicious apples* in the three main producing states: Durango, Chihuahua and Coahuila in northern Mexico, and secondly, ***what is the amount of water*** used in commercial scale production per kg of physical product, as well as the amount of water used per USD of profit as well as the amount of water associated with the creation of a job in Red Delicious apple production in the three states indicated in northern Mexico in the agricultural year 2022.

2.2 Hypothesis

First hypothesis: ***The physical water footprint*** of apple (*Malus domestica L.*) Red Delicious variety from the three main Red Delicious apple producing states, the states of Durango, Chihuahua and Coahuila in northern Mexico, determined by a Physical Water Efficiency index "EFA" (measured in liters of water used in production per kg produced abbreviated as L kg⁻ will have a *physical* water efficiency index *lower* than the world average water footprint of 700 liters per kg of apple only in the state of Chihuahua, while Durango and Coahuila will have an EFA higher than the world average water footprint reported by Mekonnen and Hoekstra (2010)[3] of 700 liters per kg.

Second hypothesis: ***The economic water footprint*** of apple (*Malus* *domestica L.)* Red Delicious variety, determined by an Economic Water Efficiency index "EEA", measuring EEA in cubic meters of water used in production for each USD of gain (if the index is positive) or loss (if the index is negative) produced (abbreviated as m^3 USD^{-1}) and an Economic Water Productivity index "EWP" (measuring EWP in USD of gain per cubic meter of water used in production, abbreviated as USD m^{-3}) will be ***higher*** than the EEA and EAP of the weighted average of the three producing states only in the state of Chihuahua, while Durango and Coahuila will then have an EEA and EAP lower than the weighted average of the three producing states.

Third hypothesis: ***The social water footprint*** of apple (*Malus domestica L.)* Red Delicious variety, determined by a Social Water Efficiency index "SWE", measured SWE in cubic meters of water used in production per equivalent job produced (abbreviated as m^3 $Jobs^{-1}$) and a Social Water Productivity index "WSP" (measured WSP in Jobs generated per cubic hectometer of water used in production, abbreviated as E hm^{-3}) will be ***higher*** than the ESA and PSA of the weighted average of the three producing states only in the state of Chihuahua, while Durango and Coahuila will then have an ESA and PSA lower than the weighted average of the three producing states.

---

[3] ***Mekonnen, M.M. and Hoekstra, A. Y. 2010.*** The green, blue and grey water footprint of crops and derived crop products. Hydrology and Earth System Sciences, 15(5): 1577-1600.

III.   LITERATURE REVIEW

3.1 Genetic origin of the apple (*Malus domestica L.)*, goodness and characteristics.

According to Botanical- Online (2023) "Apples are the fruit of apple trees, trees (*Malus spp*) of the Rosaceae family, which includes wild or cultivated shrubs such as blackthorn, roses, or other trees such as almond or cherry. The cultivated European apple tree corresponds to the species *Malus domestica L.* It is a deciduous tree up to 15 m tall. Its stems are gray and its young branches are pubescent. Elliptic-ovate leaves with the underside covered with fuzz, toothed, up to 15 cm long. White or pink flowers of up to 5 cm. The fruit (apple) is a knob of more than 5 cm, of very variable color according to the varieties. It is native to the East, formed by hybridization of wild species and sometimes feral. It is not known exactly which is the first cultivated species from which all the others derive, although it is thought most likely to come from the species *Malus sieversii Ledeb,* whose origin must be placed 15,000 or 20,000 years ago in central Asia, specifically in Tian-Shan, a mountainous area located northwest of China and Kazakhstan." [4]

According to Haro and Moreau (2023)[5] , the apple has the following health properties: "Ideal as a mid-morning or mid-afternoon snack, the apple is one of the healthiest fruits known and, as such, one of the most recommended to include in the diet. Natural remedies, handed down

---

[4] **Botanical-oline,**   2023. apples,   fruit.   availableat   : https://www.botanical- online.com/botanical-oline/apple-characteristics#Apple_characteristics.

[5] **Haro, G. Ana and Moreau, B. Mría Del Carmen. 2023.** Apple. In: PULEVA; Bienestar para disfrutar de la vida. Available at. https://www.lechepuleva.es/aprende-a-cuidarte/tu-alimentacion- de-la-a-z/m/manzana#:~:text=La%20manzana%20nos%20aporta%20vitaminas,Para%20forfortalecer%20pelo%20y%20y%20u%C3%B1as.

from generation to generation, qualify it as ideal in case of lung ailments, mixed with a spoonful of honey to cut coughs, for oral hygiene of the gums, in case of physical and intellectual exhaustion, in case of gout, as an excellent antidiarrheal.... How do apples help us?", Haros and Moreau (2023 *Op. Cit.*) point out that the nutritional and health properties of apples can be summarized in the following sixteen points:

1)   **"To appease anxiety**. When you can not concentrate because you have an empty stomach, it is advisable to take an apple, as it only provides 50 Kcal per 100 grams. This is going to avoid other more caloric foods, such as potato chips, pastries, chocolate or sweets, and which are unhealthy, since, among other things, they provide more calories that are going to be stored as fat."

2)   **"As a natural dentifrice**. The fibers of the apple (pink parts of both the skin and the pulp), combined with the force of chewing, have a dragging and cleansing effect on food residues in the mouth, as well as strengthening the gums. Being an aromatic fruit, it also acts against bad breath".

3)   **"In the fight against cholesterol**. The apple contains a fiber, pectin, which is ideal for reducing cholesterol ingested in the diet. It carries it away before it is absorbed by the body. It is the perfect fruit to complement a meat or egg dish. Also its fiber is depurative on an empty stomach because it reduces bile cholesterol."

4)   **"To improve memory.** The apple provides us with vitamins B1 and B6, which prevent mental exhaustion and strengthen the memory. It is also a source of phosphorus, a mineral present in the phospholipids of the brain, potassium and sodium, essential for nerve conduction."

5)   **"To strengthen hair and nails**. The apple provides iron, essential for strong hair and nails. It also contains pantothenic acid or vitamin B5, which is a hair regeneration enhancer."

6) **"Against obesity**. Fresh fruit, juices or some low-calorie whole-grain cookie should be the complementary food that satiates the excessive appetite of young people, because of the few kilocalories they provide, if we do not want them to walk towards overweight and obesity. Of course, all this complemented with the practice of some sport on a regular basis."

7) **"As a reinforcement for the defenses.** The vitamin C contained in the apple stimulates the immune cells, reinforcing the defenses and preventing us from colds and flus."

8) **"As an ally against hemorrhages.** Its vitamin C strengthens the walls of blood vessels and helps in the healing dwounds, by stimulating the formation of collagen. It also prevents spontaneous bleeding from nose and gums."

9) **"As an antidote against decay**. The apple is the fruit with the highest fructose content that exists. This sugar is a monosaccharide, which is a simple molecule of immediate use by the organism, which takes us out of a drop in blood glucose, avoiding the typical tiredness and dizziness that can appear between one meal and another."

10) **"As an aid to growth**. The apple is a source of calcium and phosphorus, essential in the formation of bone mineral salts. It also provides vitamin C, which is involved in the formation of the bone matrix substance. In addition, B vitamins are necessary for the growth and development of muscles. For all these reasons, it is good to eat apples during the growing season."

11) **"To increase muscle capacity.** Vitamin B1 or thiamine prevents muscle fatigue, while vitamin B2 helps in obtaining energy and vitamin B6 intervenes in the proteins that form muscle mass."

12) **"To rest the eyes**. When our eyes dry out because of poorly ventilated places, heating, computer use or simply by wearing contact lenses for a long time, apples provide water with vitamin B2 that moisturizes and improves the mechanism of vision. Apples are especially indicated when wearing contact lenses."

13)   **"Against acne**. To avoid acne, it is necessary to have a hydrated and purified skin; in this sense, the apple is a good source of water (85%). Similarly, as it does not contain fat or refined sugars, it does not contribute to cause it."

14)   **"As an isotonic drink**. The body's cells must have an adequate level of hydration, since the human body is made up of 70% water. The kidneys filter 2.5 liters of water daily, which must be replenished with 1.5 liters of pure water and a further 1.5 liters from food. The apple is 85% water and contains dissolved vitamins, minerals and sugars; therefore, this fruit quenches thirst and maintains the water level in all the cells of the body. In addition, it is easy to transport and can be taken at any time of the day."

15)   **"As an intestinal regulator**. The apple contains in its pulp substances called pectins that effectively regulate intestinal transit. Because of their ability to swell with water, they facilitate peristalsis, without interfering with the absorption of calcium, iron or magnesium, but blocking the passage of cholesterol to the cells of the intestinal surface. When you want to use it to avoid constipation, you should eat the apple with the peel. On the other hand, to avoid diarrhea it should be eaten without the peel, and it is more advisable to take it grated."

16)   **"Against the harmful effects of pollution and tobacco.** Fumes from pollution and tobacco oxidize the body's cells. The amount of vitamin C provided by the apple has an antioxidant effect on the tissues, preventing premature aging."

Because of its high nutritional value, this fruit is the most consumed in the world, because for every 100 grams, the nutritional value of the apple is characterized by 52 calories, 85.6 grams of water,

0.2 g of total fat, 0.0 mg of cholesterol, 1 mg of sodium, 107 mg of potassium, 14 g of carbohydrates and o.3 g of proteins, as well as minerals such as copper, manganese, phosphorus and zinc, as well as vitamins B, B12, folate and choline[6] .

As for apple taxonomy, according to Cerdán and Suárez (2020)[7] , apple belongs to the Order Rosales, Family Rosaceae, Subfamily Amygdoaloideae, Tribe Maleae, Genus Malus species M. domestica (Borkh, 1803.).

Based on Cerdán and Suárez (2020, *Op. Cit.),* there are several centers of origin of cultivated plants, the oldest center of origin is China, from where the apple tree originates, as well as the cherry, peach, sesame, millet, soybean, sugar cane, rice, and other crops. This center of Chinese origin would correspond to Region III that the Russian botanist Vavilov characterized for what is currently the northwest of the People's Republic of China and the north of Kazakhstan.

3.2   Global, national and regional apple production market at 2021/22.

According to FAO (2021)[8] in 2018 a total of 868 million tons of different fruits were produced in the world.

As part of the 868 million tons of fruit produced in the world, the apple, in

---

[6] **Source: https://www.herbazest.com/es/hierbas/manzana**

[7] **Cerdán, Carlos and Suárez, Ana Isabel. 2020**. Biodiversity Centers of Origin. Available at: . https://www.uv.mx/personal/asuarez/files/2020/05/Semana-5-Sesion-1-Centros-de-origen-2020.pdf

[8] *FAOSTAT, 2021*.   International Year of Fruits and Vegetables   . *Available at:* https://www.fao.org/3/cb2395es/cb2395es.pdf

terms of physical volume, according to FAO (2021 *Op. Cit.*), is the fourth most produced fruit in the world, the first place is occupied by bananas with 155 million tons, the second place was occupied by citrus fruits such as orange, lemon, grapefruit with 152 million tons, the third place was for watermelon and melon that had a volume of 131 million tons, the fourth position was precisely for apple, pear, quince and pome with 111 million tons (see figure 1).

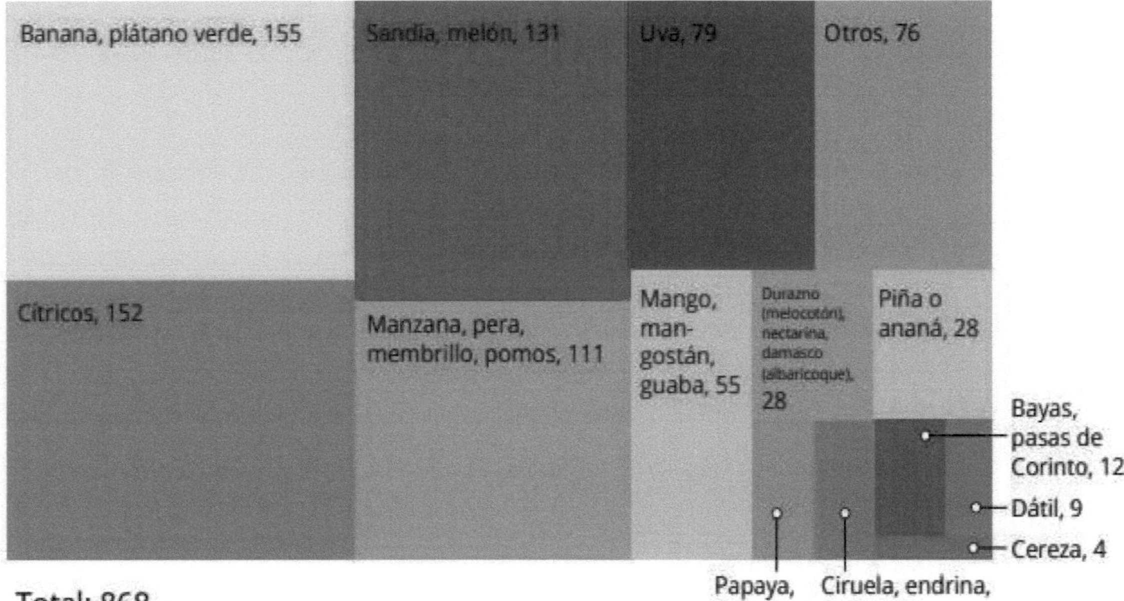

Figure 1: World fruit production by product in 2018 in million tons (Source FAOSTAT, 2021. International year of fruits and vegetables. Available at: https://www.fao.org/3/cb2395es/cb2395es.pdf)

Fruits such as grapes, mango, guava, peach, pineapple, date, papaya, plum, according to FAO (2021 *Op. Cit*) had lower volumes of physical production than apples, in fact, the production of grapes, in fifth place of importance, with 79 million tons, had a tonnage far distant from that of apples, pears, quinces and pomes (see Figure 1).

Table 1 shows the production of apples only, excluding pears, quinces and pomes as mentioned in the previous paragraphs. That source shows that between 2019 to 2021 the main apple producing countries were China, Turkey, the U.S.A., Poland and India.

Country	2021	2020	2019
China	45,983	44,066	42,425
Turkey	4493	4300	3618
United States	4467	4665	5028
Poland	4067	3555	3080
India	2276	2814	2316
World	93,144	90,490	87,509

Table 1: Main apple producing countries in the world 2019 to 2021 (in thousands of metric tons)[9]

Table 1 shows that between 2019, 2020 and 2021, world apple production increased from 87,509 to 90,490 and from there to 93,144 million metric tons (mtm) respectively, which in terms of annual growth rate (AGR) is equivalent to growing 2.1% each year, noting that China was the main apple producing country, registering 42,425, 44,066 and 45,983 mtm respectively in those three years, which in terms of annual growth rate (AGR) is equivalent to growing 2.1% each year, noting that China was the main apple producing country, registering 42,425, 44,066 and 45,983 mtm respectively in those three years, which in terms of

[9] **FAOSTAT, 2023**. FAO Statistics Division. Available at. https://www.fao.org/3/cb2395es/cb2395es.pdf: accessed November 8, 2023.

annual growth rate (AGR) is equivalent to growing 2.1% each year.

percentage accounted for 48.5%, 48.7% and 49.4% respectively of world apple production, with a TAC of 2.7% between 2019 and 2021, above the world average annual growth of 2.1%, although with production much lower than that of China, Turkey was the second largest apple producing country, since, with the volumes indicated in Table 1, Turkey contributed 4.1%, 4.8% and 4.8% of world production, however, Turkish apple production grew at an annual rate of 7.5% going from 3,618 mtm in 2019 to 4,493 mtm in 2021, similar to Poland, which was the fourth producing country with volumes relatively close to those of Turkey (see Table 1) but its annual growth was the highest within the top five apple producing countries, since going from 3,080 to 4,067 mtm between 2019 and 2021, it had a TAC in the order of 9.7%.

The United States of America was the third largest producer, contributing between 2019 and 2021 5.7%, 5.2% and 4.8% (equivalent to 5 028, 4 665 and 4 467 mtm), India was the fifth largest apple producing country, having a decreasing growth in the period, it presented a negative TAC of 0.6% (see table 1).

The global apple market, now viewed from an export-import perspective, is shown in Table 2.

Table 2 shows that in 2021, the export-import market for apples, pears and quinces amounted to $10,787 million US dollars (MUSD), i.e., between the three fruits, from the perspective of the countries that sold apples, pears and quinces to other countries, this amount of $10,787 MUSD was collected.

Table 2: Global apple, pear and quince market in 2021. Top ten countries.

Exports				Imports			
Position	THUSD		% of participation	Position	Country	THUSD	% of participation
1	China	$1,986	18.4	$ 1	Germany	$786	7.29
2	Italy	$1,161	10.8	2	Indonesia	$661	6.12
3	USA	$949	8.8	3	Russia	$534	4.95
4	Chile	$818	7.58	4	eino Unid	$517	4.79
5	New Zealand	$733	6.8	5	India	$388	3.59
6	South Africa	$732	6.79	6	aises Bajo	$359	3.33
7	Netherlands	$561	5.2	7	Mexico	$329	3.05
8	Poland	$497	4.61	8	Vietnam	$358	3.32
9	France	$429	3.98	9	Thailand	$325	3.02
10	Belgium	$402	3.73	10	France	$310	2.88
	Rest of the world	$2,519	23.31	of the world		$6,220	57.66
	Total	$10,787	100	Total	Total $	10,787	100
	Mexico	$0.507	0.0047				

Table 2 shows that the main seller (exporter) of these three fruits was China, which with $1,986 MUSD contributed 18.4% of the value of world exports, the second place went to Italy, with 10.8% (1,161 MUSD) of the total value of apple, pear and quince exports, the U.S.A. was the third seller of these three fruits, with 949 MUSD equivalent to 8.8% of global exports of these three fruits. Mexico does not figure among the top ten exporting countries.

0.507 MUSD contributed 0.0047% of the value of apple, pear and quince exports.

Now, who were the main countries that bought the apples exported by the countries indicated on the left side of Table 2, the answer is on the right side of Table 2, from there you can see the main importing countries of apple, pear and quince, which together imported $10,787 MUSD, highlighting that the main buying country of these fruits was Germany, which with $786 MUSD contributed with 7.29% of the world imports.29% of world imports, the second main importing country of these fruits was Indonesia with $661 MUSD, accounting for 6.12 of the world value of imports of these three fruits, the third main purchasing country of apple, pear and quince was Russia with $534 MUSD in 2021, equivalent to 4.95 of the total value of imports.

Mexico, according to Table 2, was the seventh main importer of apples, pears and quinces, since its imports amounted to US$329 million, equivalent to 3.05% of the global value of imports. It should be noted that France sold these fruits and collected a total of US$429 MUSD while buying these three fruits from the rest of the world for a total of US$310 MUSD.

At a lower level of aggregation, only at the level of the United Mexican

States, Table 3 contains information on apple production at four levels of aggregation:

1)   Apple production in general, i.e. all varieties, both *irrigated and rainfed*, in all types of technology (open field and protected agriculture (greenhouse, shade netting and macro-tunnel), all types of production (conventional and organic), for the domestic and foreign markets and at the state level (national, Chihuahua, Coahuila, Durango and the three producing states) (see upper part of Table 3).

2)   Apple production in general, i.e. all varieties, *only under irrigation*, in all types of technology (open field and protected agriculture (greenhouse, shade netting and macro-tunnel), all types of production (conventional and organic), for the domestic and foreign markets and at the state level (national, Chihuahua, Coahuila, Durango and the three producing states) (see upper middle part of Table 3).

3)   **Red Delicious** apple production *under irrigation*, in all types of technology (open field and protected agriculture (greenhouse, shade netting and macro-tunnel), all types of production (conventional and organic), for the domestic and foreign markets and at the state level for each of the three main producing states as well as at the national level (see lower middle part of Table 3).

4)   Production of **Red Delicious** apples **under irrigation**, all types of production (conventional and organic), for the domestic and foreign markets and **at the level of the main producing municipalities** of the three main producing states (Canatlán in open orchards, Arteaga, Coahuila in national protected agriculture and Cuauhtémoc, Chihuahua in open orchards). See bottom of Table 3.

Table 3: Apple production in Mexico, 2022

Aggregation level	Surface	production	VBP	% with respect to national		
	Ha harvested	Tons	Tons Thousands of MX$	Area	production	VBP
I) Apple production (all types of technology, IRRIGATION+TEMPORARY, all varieties, market, etc.) internal+external market)						
National	54,949.96	808,906.03	$8,871,591	100.0%	100.0%	100.0%
Durango	5,040.98	6,149.04$94	,818	9.2%	0.8%	1.1%
Coahuila	5,727.00	45,051.26$	467,736	10.4%	5.6%	5.3%
Chihuahua	31,682.36	688,788.72	$7,740,114	57.7%	85.2%	87.2%
The three states	42,450.34	739,989.02$ 8,302,667.39		77.3%	91.5%	93.6%
% of the 3 states	77.3%	91.5%	93.6%			
II) Apple production (all types of technology, irrigation, all varieties, domestic market + foreign market).						
National	41,391.68	738,974.65	$8,314,048	75.3%	91.4%	93.7%
Durango	4,372.98	5,420.44$86	,388	8.0%	0.7%	1.0%
Coahuila	4,392.00	38,846.21	$431,589	8.0%	4.8%	4.9%
Chihuahua	31,248.29	687,508.25	$7,728,442	56.9%	85.0%	87.1%
The three states	40,013.27	731,774.90	$8,246,419	72.8%	90.5%	93.0%
% of the 3 states	96.7%	99.0%	99.2%			
III) Apple production (all types of technology, irrigation, RED DELICIOUS variety, market, etc.) internal+external market). STATEWIDE						
National	13,962.79	239,903.13	$2,909,806	25.4%	29.7%	32.8%
Durango	3,238.23	4,082.42$64	,984	5.9%	0.5%	0.7%
Coahuila	460.00	3,726.00	$40,285	0.8%	0.5%	0.5%
Chihuahua	9,964.51	230,451.24$	2,790,505	18.1%	28.5%	31.5%
The three states	13,662.74	238,259.66	$2,895,773	24.9%	29.5%	32.6%
% of the 3 states	97.9%	99.3%	99.5%			
IV) Apple production (all types of technology, irrigation, RED DELICIOUS variety, domestic market + foreign market) At the level of only the three main producing municipalities. From Canatlan and Cuahtémoc is open field (CA), not protected agriculture (PA), and Arteaga is						
Canatlan, Durango CA	2,213.00	2,998.00$47	,968	4.0%	0.4%	0.5%
Arteaga, Coahuila AP	460	3,726.00	$40,285	0.8%	0.5%	0.5%
Cuauhtemoc, Chihuahua	5,344.30	124,703.20	$1,428,382	9.7%	15.4%	16.1%
The three municipalities	8,017.30	131,427.20	$1,516,634	14.6%	16.2%	17.1%
% of the 3 states	14.6%	16.2%	17.1%			

Source: Own elaboration, based on SIAP figures, 2023.

From the perspective of the first level of aggregation, Table 3 shows that in 2022 ***the rainfed apple together with the irrigated one***, was harvested in 54,949.96 ha, from which 808,906.03 tons of apples were obtained in its different varieties, production that had a value in the market equal to MX$8 871,591 thousand, noting that of these three macroeconomic figures, the main apple producing state was Chihuahua, which concentrated 57.7% of the national harvested area, contributed 852 kg of each ton of apple produced in Mexico and finally contributed with MX$872 of each MX$1000 produced by the crop in the country, and very far followed, as the second producer state, the state of Coahuila, which contributed 10.4% of the national area, contributed with 56 kg of each ton of apple produced in Mexico and contributed MX$ 53 of each MX$ produced in value for the crop at national level; Durango, although it was the third main apple producing state in Mexico, was far behind Chihuahua in percentage terms, since, from the figures of harvested area, production and value indicated at the beginning of the paragraph, it concentrated 9.The state of Durango was the third largest apple producing state in Mexico in terms of area harvested, production and value, with a production of 9.2% of the national apple harvest, 8 out of every 1000 kg of apples in Mexico were produced by Durango, and finally, out of every MX$1000 of value produced by the crop, it contributed MX$11.

At a lower level of aggregation than the one indicated in the previous paragraph, the second level of aggregation, characterized for being only *apple production surface under irrigation,* but still at the level of all the apple varieties produced, it is observed from table 3, that at national level 41,301.68 hectares (equivalent to 75.3% of the national surface indicated in the previous paragraph), from which a production of 738,974.65 tons was extracted (91.4% of the national production indicated in the previous paragraph) with a market value of MX$8,314,048 thousand (93.7% of the national value indicated in aggregation level 1). In this second level of aggregation, the state of Chihuahua continues being the main apple producing state under irrigation, since it concentrated 56.9%, 85% and 87.1% of the national harvested area, national production and national value of the crop indicated in aggregation level 1; Coahuila continued occupying the second place as apple producing state, but now only under irrigation and no longer including rainfed, since it contributed 8.0%, 4.8% and 4.9% respectively of the area, production and value of production at the national level indicated in aggregation level 1, irrigated apple in Durango represented 8.0% of the national harvested area, but its production represented only 0.7% of national production and the value of this irrigated apple represented only 1% of the value of national production.

In the third level of aggregation of Table 3, characterized by being only the *irrigated apple variety Red Delicious at state level,* (i.e. the remaining apple varieties were no longer included as in the first two levels of aggregation) it was determined that at national level 13,962.79 ha of Red Delicious apples were harvested, an area that produced 239,903.13 tons, which had a value of MX2,909,805 thousand, which in relation to the corresponding figures of aggregation level 1 (all varieties, rainfed and irrigated) represented 25.4%, 29.7% and 32.8% respectively, that is, of all the national area harvested in all apple varieties, both irrigated and rainfed, the Red Delicious apple occupied just over a quarter of the land, generated almost a third (29.7%) of the production and contributed a third (32.8%) of each peso of value that the crop produced at the national level. Once again, the state of Chihuahua continued to be the main producer of Red Delicious apples, since this variety occupied 18.1% of the national area of the crop, represented 28.5% of the overall national production and contributed 31.5 cents of each peso of value produced by the crop at the national level, but Coahuila was no longer the second but the third producer state, since Durango contributed more than Coahuila to the national production. Durango's Red Delicious variety represented 5.8% of the national apple harvested area (Coahuila's Red Delicious only 0.8%), contributed 0.5% of the national production (the same as Coahuila), but in terms of contributions to the value generated by the Red Delicious in relation to the figures of level 1 aggregation, Durango's Red Delicious apple represented MX$7 of each MX$1000 and Coahuila only MX$5.

The fourth level of aggregation in Table 3 is at the level of the three main municipalities of these three main states producing **Red Delicious apples under irrigation**. Thus, it is observed that Cuauhtémoc was the main municipality in the state of Chihuahua, with 8,017.30 ha harvested of Red Delicious apples in the open air (there was no production in protected agriculture), from which 124,703.20 tons with a value of MX$1,428,382 thousand, while Coahuila harvested this apple variety in the municipality of Arteaga, where 460 ha were harvested in protected agriculture (there was no open field production) that produced 3,726 tons with a value of MX$40,285 thousand, At the same time, in Durango, the main Red Delicious apple producing municipality was Canatlán (open field, no protected agriculture production), with 2,213 ha, which produced 2,998 tons with a market value of MX$47,968 thousand. Who was the second and third producer municipality (Arteaga or Canatlán) depends on whether we decide to visualize it from the harvested area (where Canatlán would be the second and Arteaga the third), or from the perspective of production (where Arteaga would be the second main producer and Canatlán the third) or from the perspective of value generated (again Canatlán would be the second place and Arteaga the third), see Table 3.

At the same fourth level of aggregation, the three municipalities producing the Red Delicious apple variety under irrigation together represented 14.6% of the national apple harvested area (all irrigated and rainfed varieties), the Red Delicious apple of these three municipalities contributed 16.2% of the national production and 17.1% of the value generated by the apple crop, which suggests that the Red Delicious apple variety was the main apple variety among the thirteen varieties registered by SIAP in its Statistical Yearbook of Agricultural Production portal.

3.3 Water footprint through water productivity and efficiency indexes used in the production of different crops .

The fact that the word "Red Delicious apple" appears in the title of this work would seem to suggest that the objective of this study is to study the Red Delicious apple variety, and the truth is that it is not; this work is really about how an infinitely scarce resource on earth is used: available fresh water, crystallizing the use of this scarce resource into one of an infinite number of possibilities in which water could be used: Red Delicious apple production in three northern states of Mexico in the agricultural year of 2022, so that once we have indexes about how it is that in the Red Delicious apple variety delimited the study to that geographical region and to that specific time to use them by contrasting them, with other human activities where water is used, to thus have elements of clarity that indicate in a concrete way where water is used more efficiently and more productively, in order to, if possible, assign this scarce resource, water, to those activities where it is used in the best possible way, thus contributing to sustainability and to generate the basis for fresh water to be available for future generations.

The water footprint is a concept generated at the beginning of this millennium. The concept of "water footprint" was created in 2002 by Arjen Hoekstra[10] at the UNESCO-IHE Institute for Water Education. The concept was subsequently refined and improved in its calculation methodology and accounting through a series of publications by Ashok Kumar Chapagain and Arjen Hoekstra (2004)[11] at the UNESCO-IHE Institute for Education.

In principle, the purpose of the water footprint is to estimate the amount of water used in the production of a good under specific conditions such as place, time and production system. The authors of the water footprint have classified it according to the origin of the water used in its production. The water footprint is an environmental indicator used to determine the amount of water it costs to manufacture a product; it ___is measured in units of volume per product manufactured___[12] and is made up of three different types of water depending on the source from which the water comes, which according to Zarza (2023) are defined as follows[13] :

1)   "blue water footprint: The blue water footprint is an indicator of the consumptive use of so-called "blue water", i.e. fresh surface water or groundwater. Or, in other words, the volume of freshwater consumed from surface water (rivers, lakes and reservoirs) and groundwater (aquifers)."

---

[10] **Hoekstra, A.Y. (2003)** Virtual water trade: Proceedings of the International Expert Meeting on Virtual Water Trade, Value of Water. Research Report Series No. 12. IHE Delft, the Netherlands.

[11] **Chapagain, A. K. and Hoekstra A.Y. 2004.** Water Footprints of Nations. Volume 1: Main Report. Research Report Series No. 16.

[12] As will be seen below, when HH is measured in this way, in units of water volume per unit of product, it becomes an indicator of water use efficiency.

[13] **Zarza, Laura F. 2023**. Difference between blue footprint, green footprint and gray footprint. Available at: https://www.iagua.es/respuestas/diferencia-huella-azul-huella-verde-y-huella-gris

2)   "green water footprint: The green water footprint is an indicator of human use of so-called green water, which refers to terrestrial precipitation that is not transformed into runoff or groundwater, but is stored in the soil or remains temporarily on the surface of the soil or vegetation. Or put another way, it attends to the evaporation experienced during the processes."

3)   "gray water footprint: The gray water footprint is defined as the volume of freshwater needed to assimilate the pollutant load based on concentrations under natural conditions and existing environmental water quality standards or legislation.In short, to water that is polluted as a result of processes."

Table 4 shows the world average water footprint (WF) of some products, showing that apples have a WF of 70 liters of water per 100 grams of apple, i.e., one kg of apple has a WF of 700 liters.

Table 4: Global average water footprint of selected agricultural, livestock and industrial products.

Product	Virtual content of water (liters)	Product	Virtual water content (liters)
1 glass of beer	75	1 glass of winw (125 ml)	120
1 glass of milk (200 ml)	200	1 glass of apple juice (200 ml)	190
1 cup of coffee (125 ml)	140	1 glass of orange juice (200 ml)	170
1 cup of tea (250 ml)	35	1 bag of potato crisps (200 g)	185
1 slice of bread (30 g)	40	1 egg (40 g)	135
1 slice of bread wirh cheese (10 g)	90	1 hamburger (150 g)	2400
1 potato (100 g)	25	1 tomato (70 gr)	13
1 apple (100 g)	70	1 orange (100 g)	50
1 cotton T-shirt (medium sized 500 g)	4100	1 pair of shoes (bovine leather)	8000
1 sheet of A4-paper (80 g/m^2)	10	1 microchip (2 g)	32

Source: Chapagain, A. K. and Hoekstra A. Y. 2004. Y. 2004. Water Footprints of Nations. Volume 1: Main Report. Research Report Series No. 16

Table 4 shows that in the case of milk, which although not explicitly stated, refers to bovine milk, tewater footprint is 200 liters of water per 200 ml of milk, which is equivalent to 1000 liters of water per liter of milk; however, as mentioned above, the WF is a dependent variable that depends, among other things, on time, Thus, for example, based on Mekonnen and Hoekstra (2012)[14] the HH of pastured bovine milk is 1,191 m^3 of water per ton of milk (equivalent to 1,231 liters of water per liter of milk if a milk density of 1.034 kg per liter), while in mixed production systems, i.e., combining grazing with industrialized housing, the HH drops to 956 m^3 / liter of milk.

Since the HH depends primarily on two independent variables, the volume of water "V" used in production and the quantity of product "Q" achieved with that volume of water, the HH can then be expressed, in the first place, as Zarza (2023 *Op. Cit.*) pointed out above, as a quotient in which in the numerator is the volume "V" of water used in production while in the denominator is the amount of product "Q" achieved with that volume of water, thus, the HH would be measured with the units of m^3 per unit of product, however, these same two variables, "V" and "Q" when inverted in the quotient, so that now "Q" goes in the numerator and "V" in the denominator, allows generating an index number that measures the productivity with which water is used in production, since it would be measured with the units of product "Q" obtained per unit volume "V" of water used in production.

---

[14] **Mekonnen, M. M. & Hoekstra, A, Y. 2012**. A global Assesment of the wáter footprint of farm animal products. Ecosystems (2012) 15:401-415. DOI:10.1007/s10021-011-9517-8

Thus, based on the above, Rios and collaborators (2015[15] , 2016[16] , 2018[17]), determined that HH can be measured by an efficiency index (HH = V/Q) or by a productivity index (HH=Q/V), which although the same two independent variables are used, the connotation of each index is different, since the first one evaluates how much water is used per unit of a good, for example, 700 liters of water per kg of apple, while the second index would be determining how productive the water is, in this case 1.4285 kg per m^3 of water[18] .

It should be noted that "Q" can assume three forms:

1)   *Physical* (in tons, kg, lbs., etc.)

2)   *Economic* (in monetary units of income or profit).   profit)

3)   *Social* (number of jobs generated)

Thus, based on the above, Ríos et al., 2015 *Op. Cit.*, 2016 *Op. Cit.* and 2018 *Op. Cit. They determined the following two general models* to estimate the HH by means of indices either of productivity or of the efficiency with which water is used in production:

$$\mathrm{Pr}\,oductivity = \frac{amount\ of\ "Q"\ (\ physical,\ economic\ or\ social\ )}{volume\ "V"\ of\ water\ used\ in\ production}$$ **Equation 1.**

[15] **Rios- Flores, J. Luis, Torres M., Miriam, Castro F., Rafael, Torres M., M.A. Ruiz T. José. 2015 a**. Determination of the blue water footprint in forage crops of DR-017 Comarca Lagunera, Mexico. Rev. FCA UNCUYO, 2015. 47(1): 101-122, ISSN print 0370-4661. ISSN (oie) 1853-8665, pp.93-107. Mendoza, Argentina

[16] **Rios-Flores, José Luis, Torres M. M. and Torres M., M. A. (2016 a).** Agricultural water productivity in pecan walnut in northern Mexico. Cases: Comarca Lagunera and Delicias, Chihuahua. ISBN978-3-639-80166-8. Editorial Académica Española. Saarbrucken, Germany.

[17] **Ríos-Flores, José Luis, Rios Arredondo, Becky Elizabeth, Cantú Brito, Jesús Enrique, Rios Arredondo, Hebrián Efraín, Armendáriz Erives, Sigifredo, Chávez Rivero, José Antonio, Navarrete Molina, Cayetano & Castro Franco, Rafael.(2018).** Analysis of the physical, economic and social efficiency of water in asparagus (*Asparagus officinalis L.)* and grapes (*Vitis* vinifera) table of DR-037 Altar-Pitiquito-Caborca, Sonora, Mexico 2018. *Revista de la Facultad de Ciencias Agrarias. Universidad Nacional de Cuyo*, 50(2). ISSN print 0370-4661, ISSN (online) 1853-8665. Mendoza, Argentina.

[18] 1 kg per 1 m^3 divided by 0.7 m^3 = 1.4285 kg m^{-3}

Thus, equations 1 and 2, by expressing "Q" concretely in physical, economic or social units, will generate the six equations shown in Table 6 of the materials and methods section, where we will have three equations of physical, economic and social water productivity (PFA, PEA and PSA, respectively) as well as three equations of physical, economic and social water efficiency (EFA, EEA and ESA, respectively).

3.3.1 Productivity and efficiency *efficiency* of the water used in the production.

Table 5 contains the physical, economic and social water footprint indices for different agricultural products, mainly from the state of Chihuahua, although they are also shown for two other geographic regions. It is presented through indicators of both **_efficiency_** of water used in production (as m^3 / kg, m^3 / USD profit and m^3 / employment for physical water efficiency "EFA", economic water efficiency "EEA" and social water efficiency "ESA" respectively) and **_productivity_** of water used in production (as kg / m^3, USD profit / m^3 and Jobs / hm^3, for physical water productivity "PFA", for economic water productivity "PEA" and for social water productivity "PSA" respectively).

Table 5: Water footprint through productivity and efficiency indexes of water used in the production of some agricultural products. In blue and red the products with minimum and maximum productivity and efficiency. Brown for PFA and EFA, yellow for PEA and EEA and green for PSA and ESA.

			HH through Productivity Index of water			HH through the Efficiency Index of the water		
		LocationSource	PFA (kg/ m³)	ADP (USD/m³)	PSA (Jobs/ hm³)	EFA (m³ /kg)	EEA (m³ /USD profit)	ESA (m³/Employee)
Apple (MT all the varieties)	Chihuahua	Gamboa, Rios and Ros, 2022	2.28	$0.620	34	0.439	1.613	29,412
Walnut MT	Chihuahua	Gamboa et al, 2022	0.15	$0.130	9	6.667	7.692	111,111
Grenada	Chihuahua	Rios, Hernandez and Azpilcueta, 2022	1.61	$0.903	40.8	0.621	1.108	24,510
Peanut	Chihuahua	Amaya, Rios and Chávez, 2021	0.732	$0.187	18.9	1.366	5.361	52,910
Blackberry	Michoacán	Rios, Hernandez and Chavez, 2021	1.44	$2.170	111.46	0.694	0.461	8,972
Strawberry	Michoacán	Rios, Hernandez and Chavez, 2021	8.17	$3.590	77.42	0.122	0.279	12,917
0.213 Liters of milk/								
Bovine milk	Chihuahua	Rios, Rios y Rios, 2019	m³= 0.220 kg of milk/m³	$0.005		4.545	222.222	
Cotton	Chihuahua	Rios, Pizarro and Galván, 2021	0.358	$0.057	11.63	2.790	17.536	85,985
Watermelon	La Laguna, Mexico	Armendariz, Rios and Rodriguez, 2021	14.39	$0.560	58	0.069	1.786	17,241

Source: Own elaboration, based on figures reported by the authors. Note: in the case of ADP, some authors reported USD profit per hm³ of water, standardized to USD per m³ after dividing by one million.

34

From that source, Table 5, it is observed that the AFP index (in kg / m^3) oscillates notoriously depending on the product, varying from the lowest AFP with 0.15 kg /m^3 in the case of MT walnut (medium use of technology) produced in the state of Chihuahua according to Gamboa, Rios and Rios (2022)[19] to the highest AFP crop with an index of 14.39 kg /m^3 in watermelon produced in La Laguna, Mexico according to Armendariz, Rios and Rodriguez (2012)[20] , noting that the second and third place, within the most productive in physical terms when using water in production, were in second place strawberry from Michoacan with 8.17 kg / m^3 according to Rios, Hernández and Chávez (2021), and in third place the MT apple, in general, i.e. without dividing the production among the different apple varieties produced in Chihuahua, with an index of 2.28 kg / m^3 according to Gamboa, Rios and Rios (2022, *Op. Cit)*.

Among the least productive, according to Table 5, after the walnut already mentioned, were bovine milk from Chihuahua with 0.220 kg /m^3 determined by Rios, Rios and Rios (2019)[211] , and cotton from Chihuahua with an index of 0.358 kg / m^3 according to Rios, Pizarro and Galván (2021)[22] , and intermediate to those extremes of minimum and maximum water productivity were the crops, all from the state of Chihuahua: pomegranate (1.61 kg / m^3 , according to Rios, Hernández and Azpilcueta, 2022), blackberry (1.44 kg m^3 , according to Rios, Hernández

[19] **Gamboa, Narvaéz Mirella; Rios-Flores, Jose Luis; Rios-Arredondo Becky Elizabeth. 2022.** Efficiency and productivity of water used in pecan nut (*Carya illinoensis*) versus apple (*Malus domestica*) production in Chihuahua, Mexico. Editorial Académica Española. ISBN 978603887532. Berlin, Germany.

[20] **Armandáriz, Erives Sigifredo; Rios-Flores, J. Luis; Rodríguez Santiago Yesenia. 2021.** Rentabilité et utilisation de l'eau dans la culture de la pastèque (*Citrillus lanatus L.*) goutte à goutte à La Laguna, Mexique. EDITIONS NOTRE SAVOR. ISBN 9786203473803. Riga, Latvia

[21] **Rios-Flores, José Luis; Rios-Arredondo, Becky Elizabeth and Rios-Arredondo, Hebrián Efraín. 2019.** Physical and economic water footprints of bovine milk from Delicias, Chihuahua, Mexico. Editorial Académica Española. ISBN 9786200025180. Berlin, Germany.

[22] **Rios-Flores, José Luis; Pizarro, Quezada Alejandro and Galván, Cervantes Raúl V. 2021.** Wassersparende landwirtschaftliche Muster mit Hilfe des Wasser-Fußabdrucks fall von DR005 Delicias, Chihuahua. Verlag Unser Wissen ISBN9786204117300. Berlin, Germany

and Chávez. 2021 *Op. Cit.*)[23] and peanut (0.732 kg / m^3 , according to Amaya, Rios and Chávez, 2021).[24]

As the inverse of the physical water productivity index, the physical water efficiency (in m^3 / kg) shows the same order as indicated above for the extremes of minimum and maximum PFA, thus, the crops with the minimum and maximum efficiency of water used in production were the MT walnut (produced under conditions of medium use of "MT" technology) from Chihuahua and watermelon from La Laguna, with EFA indexes of 0.069 m^3 /kg and 6.667 m^3 /kg, respectively. In other words, producing one kg of biomass required only 69 liters of water in the case of watermelon, while that same kg of biomass, but in the form of MT nut from Chihuahua, required 6,667 liters of water. The remaining crops shown in Table 5 were intermediate.

3.3.2 The productivity and *economic* efficiency of the water used in the production.

Table 6, in its part dedicated to the economic productivity and efficiency of water used in production, is reflected in the EAP (economic productivity of water) and EEA (economic efficiency of water) variables, the former, the EAP, measured in the units of USD profit/m^3 , abbreviated as USD/m^3 , and the latter, the EEA, measured in the units of as m /USD.^{3}

In terms of water productivity, at first glance, it is observed that the order was not the same as in physical water productivity, since the extremes of minimum and maximum ADP were not the same crops, since in AFP the

---

[23] **Rios-Flores, José Luis; Hernández, Ibarra Gonzalo and Chávez, Rivero José Antonio. 2021.** Economic productivity of water in Blackberry (*Rubus spp. L.*) and strawberry (*Fragaria vesca L.*): The case of Irrigation District 061 in Zamora. OUR KNOWLEDGE PUBLISHING. ISBN9786203189537. Berlin, Germany

[24] **Amaya, Lona Victor Manuel; Rios-Flores, José Luis and Chávez, Rivero José Antonio. 2021.** Efectywnosc wykorzystania wody w produkcji orzeszków ziemnuch (*Arachis hipogaea L.*) w Chihuahua. WYDAWNICTWO NASZA WIEDZA. ISBN 9786203316445 Riga, Latvia.

minimum and maximum AFP crops were MT walnut from Chihuahua and watermelon from La Laguna respectively, while in ADP the minimum and maximum ADP crops were bovine milk from Delicias[25] , Chihuahua and strawberry crop from Michoacán[26] respectively, since their index numbers show so: 0.005 and 3.590 USD/m^3 respectively (see Table 5).

The remaining crops were left with EAP indices intermediate to the indices indicated for bovine milk and strawberries in the previous paragraph, thus, starting from the second crop with the lowest EAP to the second with the highest EAP, we had cotton (with 0.057 USD/m)[327] , pecan nut[28] (with 0.130 USD/m^3), peanut[29] (with 0.187 USD/m^3), watermelon[30] (with 0.560 USD/m^3), pomegranate[31] (with 0.903 USD/m^3), generic apple MT[32] (not separated by varieties, with 0.620 USD/m^3), and blackberry[33] was the second crop with the second highest ADP with 2.170 USD/m^3

Analyzing now the EEA (economic efficiency of water), since the efficiency function is an inverse function of the productivity function, necessarily, we had the same order of crops with minimum and maximum economic productivity indicated two paragraphs ago: that is, in the EEA

---

[25] **Rios-Flores, José Luis; Rios-Arredondo, Becky Elizabeth and Rios-Arredondo, Hebrián Efraín. 2019.** *Op. cit.*

[26] **Rios-Flores, José Luis; Hernández, Ibarra Gonzalo and Chávez, Rivero José Antonio. 2021.** *Op. cit.*

[27] **Rios-Flores, José Luis; Pizarro, Quezada Alejandro and Galván, Cervantes Raúl V. 2021.** *Op. cit.*

[28] *Gamboa, Narvaéz Mirella; Rios-Flores, Jose Luis; Rios-Arredondo Becky Elizabeth. 2022.* *Op. cit.*

[29] **Amaya, Lona Victor Manuel; Rios-Flores, José Luis and Chávez, Rivero José Antonio. 2021..** **Op. cit.**

[30] **Armandáriz, Erives Sigifredo; Rios-Flores, J. Luis; Rodríguez Santiago Yesenia. 2021..** Op. cit.

[31] **Rios-Flores José Luis, Hernández Ibarra Gonzalo and Azpilcueta Ruiz-Esparza Manuel De Jesus. 2021.** Economia agricolo.ambientale dell'acqua utilizzata sulla produzione commerciale di melograno (*Punica granatum L.*) EDIZIONI SAPIENZA. ISBN 9786205168288. Berlino, Germania.

[32] *Gamboa, Narvaéz Mirella; Rios-Flores, Jose Luis; Rios-Arredondo Becky Elizabeth. 2022.* *Op. cit.*

[33] **Rios-Flores, José Luis; Hernández, Ibarra Gonzalo and Chávez, Rivero José Antonio. 2021.** *Op. cit.*

the products with minimum and maximum PEA were bovine milk from Delicias[34] , Chihuahua and the strawberry crop from Michoacán[35] respectively, since their EEA index numbers prove it: 222.222 and 0.279 m^3 / USD and profit respectively. There is a very noticeable difference, seen in liters of water per dollar of profit, from 222,222 liters of water invested to produce one dollar of profit in the production of bovine milk to only 279 liters of water in the case of strawberries, i.e., producing one dollar of profit in the production of bovine milk has a cost in terms of water investment used in production equal to 798 times the same volume of water used by the strawberry crop (see Table 5).

The other crops in Table 5 showed an FSS intermediate to those indicated as minimum (bovine milk) and maximum (strawberry) FSS, since they had FSS indexes intermediate to the indexes indicated for bovine milk and strawberry, following the same descriptive logic used in the analysis of the intermediate FSP crops, starting from the second crop with minimum FSS to the second with maximum FSS, we had the cotton crop[36] (with 17.536 m^3 /USD gain), pecan nut[37] (with 7.692 m^3 /USD gain), peanut[38] (with 5.361 m^3 /USD gain), watermelon[39] (with 1.786 m^3 /USD gain), the generic apple MT[40] (not separated by varieties, with 1.613 m^3 /USD gain), pomegranate[41] (with 1.108 m^3 /USD gain), and

---

[34] **Rios-Flores, José Luis; Rios-Arredondo, Becky Elizabeth and Rios-Arredondo, Hebrián Efraín. 2019.** *Op. cit.*

[35] **Rios-Flores, José Luis; Hernández, Ibarra Gonzalo and Chávez, Rivero José Antonio. 2021.** *Op. cit.*

[36] **Rios-Flores, José Luis; Pizarro, Quezada Alejandro and Galván, Cervantes Raúl V. 2021.** *Op. cit.*

[37] *Gamboa, Narvaéz Mirella; Rios-Flores, Jose Luis; Rios-Arredondo Becky Elizabeth. 2022. Op. cit.*

[38] **Amaya,**

[38] **Amaya, Lona Victor Manuel; Rios-Flores, José Luis and Chávez, Rivero José Antonio. 2021.. Op. cit.**

[39] **Armandáriz, Erives Sigifredo; Rios-Flores, J. Luis; Rodríguez Santiago Yesenia. 2021.**.. Op. cit.

[40] *Gamboa, Narvaéz Mirella; Rios-Flores, Jose Luis; Rios-Arredondo Becky Elizabeth. 2022. Op. cit.*

[41] **Rios-Flores José Luis, Hernández Ibarra Gonzalo and Azpilcueta Ruiz-Esparza Manuel De**

blackberry[42] was the second crop with the second highest FSS with 0.461m^3/USD gain.

3.3.3 Productivityand efficiency *efficiency* of the water used in the production.

Although, in general, the extreme products of minimum and maximum productivity were different for both crops, the minimum and maximum productivity and efficiency, seen at a particular level, when considering the question of whether there was only one crop that concentrated the minimum productivity and efficiency in its three aspects, physical, economic and social, it is observed that the answer is the walnut crop, It is enough to see that this line is almost entirely blue, "almost" because only in productivity and economic efficiency it was not the worst crop, but in physical terms it was the one that demanded more water to produce a kg of biomass, in economic terms, it was the one that used more water to produce a USD of profit. The same did not happen in terms of whether any crop was the best in physical, economic and social terms in terms of water use in production, as there were three different crops.

The social productivity of water "WSP" used in agricultural production, as already mentioned, was measured in Table 5, by the jobs generated associated[43] with the use of one cubic hectometer of water (i.e., one million cubic meters of water, the usual measure in large water reservoirs), abbreviated as jobs /hm^3 , and the social efficiency of water

---

Jesus. 2021. *Op. cit.*

[42] **Rios-Flores, José Luis; Hernández, Ibarra Gonzalo and Chávez, Rivero José Antonio. 2021.** *Op. Cit.*

[43] The water used in production, *per se*, does not generate employment, but, as each crop has a specific water demand at the commercial level (that volume of water is equal to the multiplication of the 10 thousand m^2 of a hectare by the usual irrigation lamina of the crop at the commercial scale) as well as a specific labor demand (measured in days per hectare), water used in production is associated with the generation of different amounts of employment in each crop.

"ESA" measured as the amount of cubic meters of water that were associated with the generation of one job, in Table 5, indicates that the crops in the state of Chihuahua, starting from the crop with maximum PSA and ESA, and moving in the direction of the one with minimum PSA and ESA, had the following order: blackberry (with 111.5 E/hm^3 and 8,972 m^3 /job respectively the PSA and ESA)[44] strawberry (with 77.42 jobs/hm^3 and 12,972 m^3 /job)[45] , watermelon (with 58 jobs/hm^3 and 17,241 m^3 /job)[46] , pomegranate (with 408 jobs/hm^3 and 24,510 m^3 /job)[47] , generic apple MT (with 34 jobs/hm^3 and 29,412 m^3 /job)[48] , peanut (with 18.9 jobs/hm^3 and 52,910 m^3 /employment)[49] , cotton (with 11.63 jobs/hm^3 and 85,985 m^3 /employment)[50] and MT walnut (with 9 jobs/hm^3 and 111,111 m^3 /employment)[51]

It should be noted that in the three types of water efficiency and productivity, i.e., in physical, economic and social terms, there is a notorious variation, which could necessarily raise the question of why this great difference between the productivity and efficiency of water used in production arises, the answer is given by the independent variables "Q" and "V" in equations 1 and 2.

Since "Q" involves variables such as the physical yield "RF" per hectare, the price "p" per ton, the cost "C" per hectare and finally the exchange parity "PC", variables that are very specific, very inherent to each crop, suffice it to say, just as an example, that if a crop has a "large" RF, let us say alfalfa, with a RF of 80 ton/ha and another has a RF of only 0.9

---

⁴⁴ **Rios, Hernández and Chávez, 2021.** *Op. cit.*
⁴⁵ **Rios, Hernández and Chávez, 2021.** *Op. cit.*
⁴⁶ **Armendáriz, Rios and Rodríguez, 2021.** *Op. cit.*
⁴⁷ **Rios, Hernandez and Azpilcueta, 2022**
⁴⁸ **Gamboa, Rios and Rios, 2022.** *Op. cit.*
⁴⁹ **Amaya, Rios and Chávez, 2021. Op. cit.**
⁵⁰ **Rios, Pizarro and Galván, 2021.**
⁵¹ **Gamboa, Rios and Rios, 2022.** *Op. cit.*

ton/ha, this will necessarily have an impact on the physical productivity of water, the same with "p" and "C", moreover, suppose we are contrasting the same crop, suppose red tomato, but one is irrigated by gravity-rolled water and the other with groundwater irrigation by drip irrigation, necessarily even if it is the same crop, they will have different cost per hectare, and possibly even different prices, this will necessarily imply a different economic efficiency of the water used in production, on the other hand, the variable "V", i.e. the volume of water used per hectare, will depend on the irrigation lamina "LR", which is different in each crop, for example, there will be a crop whose LR is a water column of 2.5 meters high while for another crop a LR of 0.5 will be enough, moreover, even if it is the same crop but in a different irrigation system, if the irrigation is not very technified with water irrigated from some distant dam by gravity, possibly its efficiency index of conduction of the hydraulic network "EC" conducting water is low, say 0.5 (i.e., for every liter that is released in the dam, only half a liter of that water will reach the crop, since the rest will be "lost" by evaporation, infiltration, etc.), but if the water source, the dam, is close to the plot, even if it is by gravity, the fact that it is close will make the EC be higher than 0.5, or eenmore, if the crop is irrigated with groundwater by micro-drip, there will be few water losses and EC will tend to unity, that is, to take advantage of the entire volume of water released from the water source to irrigate the crop.

IV. MATERIALS AND METHODS

4.1 Location of the area of study

The three main apple producing states in Mexico analyzed in this study, Chihuahua, Coahuila and Durango, are located in the north of the United Mexican States, seen as a single joint geographical area, bordering the United States of America to the north, to the south with the states of Nayarit and Zacatecas, to the east with the state of Nuevo León and to the west with the states of Sonora and Sinaloa (see Figure 2).

Figure 2: Geographical location of the three main apple producing states in Mexico: Chihuahua, Coahuila and Durango.

Although the three main apple-producing states in Mexico have differences in climate, soil, vegetation, etc., in reality these differences are not so marked, since the three states belong to the eco-geographic region known as the Chihuahuan Desert, which includes part of northern Mexico where Chihuahua, Coahuila and Durango are located, as well as part of southern Texas in the USA.

The three states are arid regions, with low rainfall ranging from 240 to 600 mm per year. They also have mountain ranges with cold climates, which is where apple production is located, where various species of oaks, pines and firs can be found.

The main municipalities producing Red Delicious apples in the three states analyzed are the municipalities of Cuauhtémoc, in Chihuahua, Arteaga, in Coahuila, and Canatlán, in the state of Durango.

4.2 Mathematical models used in the calculation of the physical, economic and social water footprint and the sources from which they were fed

The water footprint in its physical, economic and social forms will be estimated using the mathematical models shown in Table 6, source where the equations that allow estimating the physical (PFA), economic (PEA) and social (PSA) productivity of water used in production appear, as well as the inverse equations, which result in the mathematical functions that allow calculating the physical (EFA), economic (EEA) and social (ESA) efficiency of water used in production. These models generated by Ríos *et al* (2015 *Op. Cit.*, 2017 *Op. Cit.* and 2018 *Op. Cit.*) have been widely used in the calculation of water efficiency and productivity in multiple crops, as well as in livestock activities such as milk production.

The meaning of each of the independent variables on which the PFA, PEA, PSA (initials for Physical, Economic and Social Productivity of Water used in production), EFA, EEA and ESA (initials for Physical, Economic and Social Efficiency of Water used in production) depend, as well as the source from which they were taken, are as follows:

RF_i = Physical yield of the i-th crop (in ton ha^{-1}). Where RF =GVP/P; the sources of GVP (Gross Value of Production and S = harvested area (in ha) are sourced from SIAP-SADER (2023). "Cierre agrícola2022   ."
Available at https://nube.siap.gob.mx/cierreagricola/

LR_i = irrigated area of the i-th crop (in m). Source: INIFAP- INIFAP-CENID-RASPA, 2006.[52]

EC_i = Hydraulic conduction efficiency of the i-th crop. 0 ' EC
1 (Source: INIFAP.CENID-RASPA, 2006 *Op. Cit.*).

S_i = Harvested area of the i-th crop (in ha). Source: SIAP- SADER (2019). Source: SIAP-SADER (2023 *Op. Cit.*).

p_i = Price of the product of the i-th crop (in MX\$ ton^{-1}). Where pi= VBPi/Pi, the sources of VBP (Gross Value of Production and P = annual physical production (in tons) are from SIAP- SADER (2023 *Op. Cit.*). PC = Exchange parity, Mexican pesos (MX\$) per USD. Source: Currency converter: available at:
https://www.xe.com/es/currencyconverter/convert/?Amount=1&From= USD&To=MXN.

---

[52] **INIFAP-CENID-RASPA. (2006).** *Irrigation Program.* [Date of consultation: December 1, 2023]. Available at: https://cenidraspa.org/serg/serg_v1.php

C_i = Cost of production per hectare of the i-th crop (in MX\$ $^{-1}$). Source: FIRA (2023, "Agrocost", available at: https://www.fira.gob.mx/Nd/Agrocostos.jsp).

Table 6: Mathematical models used to measure the physical (PFA), economic (PEA) and social (PSA) productivity of water used in production.

Variable	Model for a single crop	Model for a bulk crop aggregate
1. EFA (enY L kg-1)	$=10^4 * LRi * (RFi * ECi)^{-1}$	$y = \dfrac{10^4 \sum_{i=1}^{n} S_i \, LR_i \, (EC_i)^{-1}}{\sum_{i=1}^{n} S_i \, RF_i}$
2. PFA (kg m3)	$Y = 10^{-1} * RFi * ECi * LRi^{-1}$	$y = \dfrac{10^{-1} \sum_{i=1}^{n} S_i \, RF_i}{\sum_{i=1}^{n} S_i \, LR_i \, (EC_i)^{-1}}$
3. EEA (m^3 USD profit)[1]	$y = \dfrac{10^4 \left(\dfrac{LR_i}{EC_i} \right)}{RF_i \left(\dfrac{P_i}{PC} \right) - \left(\dfrac{C_i}{PC} \right)}$	$y = \dfrac{10^4 \sum_{i=1}^{n} S_i \, LR_i \, (EC_i)^{-1}}{\sum_{i=1}^{n} S_i \, ((RF_i \, P_i - C_i) / PC)}$
4. PEA (thousands of USD of gain hm-3)	$y = 10^2 \, S_i \, EC_i \, (LR_i)^{-1}$	$y = \dfrac{10^2 \sum_{i=1}^{n} S_i \, g_i}{\sum_{i=1}^{n} S_i \, LR_i \, (EC_i)^{-1}}$
5. PSA (Jobs hm-3)	$y = \dfrac{S_i \, J}{72 \, (LR_i / EC_i)}$	$y = \dfrac{\sum_{i=1}^{n} S_i \, J}{72 \sum_{i=1}^{n} S_i \, (LR_i / EC_i)}$
6. ESA (m^3 employment)$^{-1}$	$V = \dfrac{2.88 * 10^6 \, LR_i}{J \, EC_i}$	$y = \left(2.88 * 10^6 \right) \dfrac{\sum_{i=1}^{n} S_i \, (LR_i / EC_i)}{\sum_{i=1}^{n} S_i \, J_i}$

Source: Own elaboration, based on mathematical models estimators of PFA, PEA, PSA, EFA, EEA and ESA for a crop individually or for crop aggregates from Rios *et al* (2015 a) and Rios *et al* (2018).

gi= Ui = Profit per hectare of the i-th crop (in US$ ha) (in US$)[-1] =RFi(pi/PC)-(Ci/PC).

Ji = Number of labor days invested per hectare in the i-th crop.

Sources: FIRA (2022, "Agrocost", available at:

https://www.fira.gob.mx/Nd/Agrocostos.jsp)

i = i-th crop under a specific form of irrigation (pumping, gravity, drip irrigation, sprinkler irrigation, etc).

288 = Number of working days per year per employee = 6 working days per week, times 48 weeks per year.

4.3   Geographical delimitation, and meaning of abbreviations

Data on harvested area, production   "Q", VBP, physical yield per hectare "RF" and average rural price per ton of the red apple crop Red Delicious variety, considered in this study, come from SIAP-SADER, 2023 (*Op. Cit.*).

B of M: Banco de México m^3 :cubic meter

hm^3 = cubic hectometer = one million cubic meters USD: United States dollar

MUSD: Millions of U.S. dollars MX$: Mexican peso

VBP : Gross Value of production (in MX$) = Q(PMR) Q : Annual physical production (tons)

PMR: Average Rural Price (in MX$ per ton) RF: physical yield per hectare (in ton ha $)^{-1}$

RB/C: Benefit-Cost Ratio (= revenue per hectare/cost per hectare)

PFA = Physical productivity of the water used in production.

EFA = Physical water efficiency of water used in production PEA = Economic productivity of water used in production

EEA = economic water efficiency of water used in production PES = social productivity of water used in production

SWE = Social efficiency of the water used in production

V.   RESULTS AND DISCUSSION.

5.1 Results: profitability and productivity and efficiency of water used in apple (*Malus domestica*) Red Delicious production in the states of Durango, Coahuila and Chihuahua, Mexico at 2022.

Based on Table 7, a source that indicates the maintenance costs of a hectare of Red Delicious apple trees in production in the states of Durango, Chihuahua and Coahuila in 2022, there are three types of costs, the first, but not the most important, is the monetary cost, The second is the cost that indicates the amount of labor invested, both in labor days per hectare and in equivalent jobs per hectare, and the third cost, which we consider the most important, is the cost in water, that is, how much water had to be used at the commercial level to carry out the production.

Thus, the total monetary cost of production per hectare of the crop was MX\$ 95,152ha (equivalent to USD 5,556.51/ha) in Durango, MX\$ 219,681/ha (equivalent to USD 12,824.45/ha) in Chihuahua and MX\$ 362,094 (equivalent to USD 21,144.79/ha) in Coahuila, with irrigation, the most important cost item given the nature of this study, costing MX\$ 4,410/ha in Durango, MX\$ 13,533/ha in Chihuahua and MX\$ 10,080/ha in Coahuila.

The second type of cost indicated is that of labor investment. It is worth remembering here that according to Marx (1982)[53] the amount of socially necessary labor invested in a commodity is what allows it to be equated

---

[53] Marx, Carlos. 1982. Capital. Volume I. FCE, Mexico DF, 1982

and exchanged, likewise, according to David Ricardo's theory of value (1982)[54] a region, a nation, will tend to specialize and therefore, to have comparative advantages in international trade, when its commodity, in comparison with other similar or disparate commodities, enjoys a smaller investment of labor per unit. Thus, Table 7 shows that in the Durango blocks 35 workdays were invested, while in Chihuahua and Coahuila 138.96 and 198 workdays were invested, that is, 280, 1110.08 and 1584 hours of work per hectare; these amounts of work that respectively allowed obtaining the RF (ton/ha) in the locations analyzed in this work, indicated in Table 8 in the order of 1.355 ton/ha for Canatlán, Durango, 23.344 ton/ha for Cuauhtémoc, Chihuahua and 8.100 ton/ha for Arteaga, Coahuila, thus, it is easy to obtain the labor productivity index for the Red Delicious apple of Durango, Chihuahua and Coahuila: 206.7, 137.0 and 67.9 hours/ton, respectively.

The third cost item mentioned, considered by us to be the most important, is the volume of water invested in production on a commercial scale. Table 7 shows that in Durango, Chihuahua and Coahuila, respectively, it was necessary to invest a total of 8, 9 and 9 thousand m^3 of water/ha in production per hectare.

Within the total cost per hectare, there are two macro cost components: operating costs and other costs. Within the first one, the items Fertilization, Cultural work, Irrigation, Pest and disease control, and Harvest, selection and packing are consigned, which together this group of items, the operating costs, represented 80, 87 and 87% of the total cost, while the other costs (land rent, financial interests and machinery

---

and equipment depreciation) represented the missing complement of 20, 13 and 13% respectively for Durango, Chihuahua and Coahuila. See Table 7. The irrigation item implied 3, 9 and 9 irrigations respectively in Durango, Chihuahua and Coahuila.

Table 7: Maintenance production costs per hectare in apple irrigated by traditional pumping in Durango and drip irrigation in Chihu ahua and Coahuila.

	I.- Operating costs:					
	Durango		Chihuahua		Coahuila	
	Cost/ha	Days/ha		Cost/haDays/ha	Days/haCost/ha	
Fertilization	20,	788834	,8311	.5	39,	6189
Cultural work	21,	520172,	12365.	7666,	700160	
Irrigation 4,410313	,53310	,	0809			
Pest, weed and disease control	8,	231525	,7081	.	56,466	
Harvesting, sorting and packing	15,	5001835	,00070178,	48020		
Miscellaneous	5,7003		,20010		,000	
marketing 7,00050			,750			
Subtotal: 76,	149191,		395362,		094198	
Other costs:						
Ground rent	10,00010		,0003		,000	
Financial cost	6,71912,		54441		,005	
subtotal other costs:	19,00328		,28654		,868	
TOTAL (MX$/ha)	95,152	35	219,681	139	416,962	198
TOTAL (USD/ha)	5,557		12,828		24,349	
cost of (MX$/ m³) of water	0.551		.		501.12	
Volume of water (m³)used per hectare			800090009000			
U.S. dollar peso exchange rate	17.1245					
empleos/ha0			.	121527780.481805560.6875		

Source: Banco de México https://www.banxico.org.mx/tipcamb/main.do?page=tip&idioma=sp, date of consultation: November 26, 2023.

Table 8 shows the profitability of irrigated apple production, in general for all varieties, as well as in particular for the Red Delicious variety. The upper part of Table 8 shows the macroeconomic data of RF/ha, price/ton and income/ha of apples in general (all varieties, irrigated and rainfed, all types of technology, i.e. open-air production and protected agriculture in shade net, greenhouse and macro-tunnel), showing that at the national level the RF is 14.721 ton/ha, with Chihuahua being the apple producing state with the highest RF, in the order of 21,740 ton/ha, followed by the state of Coahuila with 7,866 ton/ha and even further away by the state of Durango, with 1,220 ton/ha. As for the average price "p" per ton, the national average was MX\$ 10,967, but it was not Chihuahua who had the best price, with MX\$ 11,237, the best price was Durango with MX\$ 15,420, Coahuila had an intermediate price of MX\$ 10,382. Thus, while the income "I" per hectare, as a mathematical equation is a dependent variable of RF and p, as follows: I = RF*p, then it is easy to understand why it corresponds to the state of Chihuahua to have been the state with the highest income per hectare, of MX\$ 244,304, higher than the national average of MX\$ 161,449/ha, because although it did not enjoy the best price/ton, its very high physical productivity, 21,740 ton/ha, allowed it to enjoy this high income/ha, the opposite was the state of Durango, which although it enjoyed the best price (MX\$ 15,420/ton), this was not enough to counteract the negative effect of its low RF (1.220 ton/ha), so its income/ha was the lowest of the three states. Coahuila played an intermediate position , hence its income of MX\$ 81,672/ha was intermediate to Chihuahua and Durango.

Table 8: Profitability in the cultivation of Red Delicious apples in Mexico

	Ton/ha	Price (MX$/ton)	Income (MX$)/ha	Cost (MX$)/ha	Profit (MX$)/ha	RB/C
I) Apple production (all types of technology, IRRIGATION + TEMPORARY, all varieties, domestic market+market, all types of fruit, all varieties, all types of fruit, all types of fruit, all varieties, all types of fruit, all types of fruit, all types of fruit outdoor)						
National	14.721	$ 10,967	$161,449			
Durango	1.220	$ 15,420	$18,809			
Coahuila	7.866	$ 10,382	$81,672			
Chihuahua	21.740	$ 11,237	$244,304			
The three states	17.432	$ 11,220	$195,585			
% of the 3 states						
II) Apple production (all types of technology, IRRIGATION, ALL VARIETIES, domestic and foreign market).						
National	17.853	$ 11,251	$ 200,863			
Durango	1.240	$ 15,937	$ 19,755	$ 95,152	-$ 75,398	0.21
Coahuila	8.845	$ 11,110	$ 98,267	$ 416,962	-$ 318,695	0.24
Chihuahua	22.001	$ 11,241	$ 247,324	$ 219,681	$ 27,643	1.13
The three states	18.288	$ 11,269	$ 206,092	$ 227,726	-$ 21,634	0.91
III) Apple production (all types of technology, IRRIGATION, RED DELICIOUS VARIETY, domestic market + foreign market). At STATE level.						
National	17.182	$ 12,129	$ 208,397			
Durango	1.261	$ 15,918	$ 20,068	$ 95,152	-$ 75,085	0.21
Coahuila	8.100	$ 10,812	$ 87,576	$ 416,962	-$ 329,386	0.21
Chihuahua	23.127	$ 12,109	$ 280,044	$ 219,681	$ 60,363	1.27
The three states	17.439	$ 12,154	$ 211,947	$ 196,808	$ 15,138	1.08
IV) Apple production (all types of technology (CA=open field; AP=protected agriculture), IRRIGATION, RED DELICIOUS variety, domestic market+foreign market). Conventional production (non-organic) At the level of only the three main ones producing municipalities						
Canatlan, Durango CA	1.355	$ 16,000	$ 21,676	$ 95,152	-$ 73,477	0.23
Arteaga, Coahuila AP	8.100	$ 10,812	$ 87,576	$ 416,962	-$ 329,386	0.21
Cuauhtemoc, Chihuahua CA	23.334	$ 11,454	$ 267,272	$ 219,681	$ 47,591	1.22
The three municipalities	16.393	$ 14,714	$ 189,170	$ 196,627	-$ 7,456	0.96

Source: Own elaboration based on figures in Tables 3 and 5.

Saving the same analysis as above for part II of Table 8, where the figures for apple production (all types of technology, irrigated, all varieties, for domestic and foreign markets) are presented, and pointing out only their effect on profitability, when looking at the figures for income/ha and cost/ha, we have that Durango, with an income/ha of MX$ 19,755 and a cost/ha of MX$ 95,152 had a Benefit/Cost Ratio (BCR/C) of only 0.21, that is, only 21 cents of each peso invested in this type of apple production in 2022 were recovered, incurring a loss of 79 cents of each peso invested, while in Coahuila the situation was no different, since MX$ 416,692/ha were invested but only had an income/ha of MX$ 98,267, that is, it had a BR/C of 0.The exception, apparently a good one, was Chihuahua, where the RB/C was of the order of 1.13, that is, each peso invested was recovered plus a surplus, in the form of profit, of 13 cents for each peso invested.

And it was said that it was apparently good, because although the RB/C was higher than the unit and higher than those of Coahuila and Durango, it now remains to be seen how good it was in relation to the leading rate of return in the market: the CETES rate (Certificados de la Tesorería de la Federación), which was in 2022 at an annualized level of one year, around 11.20%[55] , therefore, it did have a good RB/C, but not that good.

Let us clarify: the producers in that sphere of production, with a BR/C of 1.13 faced all the concomitant problems of production; drought, extreme rains, not gathering the necessary cold hours, pests, workers' strikes, low prices, etc., and in exchange they were rewarded with a profit of 13 cents per peso, only 1.9 percentage units above the CETES.9 percentage units above the CETES, i.e., if they had invested their money in CETES, without facing any risks such as toinherent to production, they would have obtained just 1.9 percentage units less. It is all a matter of focus.

[55] **El País, 2023.** Cetes: the financial instrument that attracts savers in times of high inflation in Mexico. Last consulted on February 9, 2023. . Available at: https://elpais.com/mexico/2023-02-11/cetes-el-instrumento-financiero-que-atrae-ahorradores-en- tiempos-de-alta-inflacion-en-mexico.html.

Part III of Table 8 is for the production of Red Delicious apples under irrigation in all types of technology, domestic and foreign market, it is observed for Durango and Coahuila a similar situation to the one indicated in the previous paragraph, since the BR/C of Durango and Coahuila was 0.21 and 0.21 respectively, not so for the Red Delicious apple from Chihuahua, where the RB/C was 1.27, much better than the CBR of 1.13 mentioned in the previous paragraph, and was not 1.9 but 15.9 percentage units above the CETES rate. The premium for the producer for assuming the risks of production were well paid by the market.

Finally, the lower part of Table 8, Part IV, for Red Delicious apple production (under irrigation, in all types of technology, for both domestic and foreign markets) in the most important Red Delicious apple producing municipalities of the three states indicated, these locations being the municipalities of Canatlán in Durango, Cuauhtémoc in Chihuahua and Arteaga in Coahuila, shows that the RB/C, in that order, was 0.23, 1.22 and 0.21, i.e., neither Canatlán nor Arteaga, as the main Red Delicious apple producing municipalities, improved their profitability, while Cuauhtémoc decreased its RB/C by 5 percentage points compared to that indicated in the previous paragraph, even so, it was above the CETES rate by almost 11 percentage units.

Finally, Table 9 contains the indicators of the water footprint "HH", the objective of this work. Vertically, Table 9 is divided into three parts, the upper part is for the physical HH, which has been measured as Y1 and Y2, Y1 is the EFA, expressed in liters of water used per kg produced, and Y2 is the PFA, expressed in kilograms of apples produced per cubic meter of water used in production, the second part shows the economic HH, evaluated as Y3 and Y4, Y3 indicates the EEA, expressed in m^3 of water used in production for each USD of gain (if positive) or each USD of loss (if negative) and Y4 was expressed as the gain in USD (if positive) or loss (if negative) generated per m^3 of water used in production.

Table 9 : Water footprint of Red Delicious apples in Mexico, 2022

Variable and units in which it is expressed	Statewide Red Delicious Apple				Red Delicious Apple a		
	Durango (CA)	Coahuila (AP)	Chihuahua (CA)	The three states	Canatlan	Arteaga	Cuauhtémo c
Physical Water Footprint: PFA = Physical Water Efficiency; PFA = Physical Water Productivity.							
$Y1$ = EFA = L/kg	6,346	1,111	389	541	5,905	1,111	386
$Y2$ = PFA = kg/m^3	0.158	0.900	2.557	1.848	0.169	0.900	2.593
Economic Water Footprint: EEA = Economic Water Efficiency; PEA = Economic Productivity. Water							
$Y3$ = EEA = m^3 of water used for each USD of gain (if positive) or loss (if negative).	- 1.82	-0.47	2.55-	39.73	-1.86	-0.47	3.24
$Y4$ = ADP = profit (+) or loss (-) (in USD) per m^3 of water used	-$0.55	-$2 .14	$ 0.39	-$ 0.03	-$ 0.54	-$2 .14	$ 0.31
Social Water Footprint: SWF = Social Water Efficiency; SWP = Social Water Productivity.							
$Y5$ = ESA = m^3 of water used by each employment generated	65,829	13,091	18,680	22,570	65,829	13,091	18,680
$Y6$ = PSA = Jobs generated by hm^3	15.19	76.39	53.53	44.31	15.19	76.39	53.53

Source: Own elaboration, based on figures in Tables 3, 7 and 8 parts III and IV. Note: CA = open field production; AP = protected agriculture production. In Coahuila all Red Delicious apple production was under open field, and in Durango and Chihuahua and Durango it was under protected agriculture.

The lower part shows the social HH, expressed as Y5 and Y6, Y5 is the ESA, expressed as m^3 of water used in production associated with the generation of one equivalent job, while Y6 is the PES, expressed as equivalent jobs generated associated with the use of one hm^3 of water.

It should be noted that water in itself does not generate employment, but when it is associated with a crop under a specific form of production with different investment of daily wages per hectare, it will therefore be associated with a greater or lesser amount of employment generated.

Thus, the Physical WF in Table 9, in its physical water efficiency index (WFE) form, shows that at the state level, Durango, Coahuila and Chihuahua, in that order, had WFE indexes equal t6346, 1111 and 389 liters of water used in production per kg of apples, noting that only Chihuahua was below, almost half of the 700 liters per kg indicated by Mekonnen and Hoeksta (2010 *Op. Cit.*) as the world average water footprint, noting that due to their high physical WF, Durango, above all, and to a lesser extent Coahuila, could hardly compete in a world market that is increasingly informed and aware of the use of scarce resources such as water, but not Chihuahua, whose low physical WF offers possibilities of opening market niches in the world for those consumers who use their purchasing power to acquire products with low WF. At the municipal level, there are no notable differences in physical WF with respect to the state averages.

Regarding the Physical WF measured as a water productivity index, Table 9 shows that the PFA, Y2, which at the state level, for Durango, Coahuila and Chihuahua, in that order, were 0.158, 0.900 and 2.557 kg/m^3 , which is consistent with the EFA, since the PFA is the inverse of the EFA. It should be noted that it is necessary to visualize the HH in both ways, since they have different connotations, since one measures the amount of water used per unit of product, while the other indicates the amount of product generated per volumetric unit of water. Thus, it can be seen that Chihuahua had the most productive use of water in physical terms, since each cubic meter of water used in the states generated in Chihuahua 16 2 times the amount of apples produced per cubic meter of water in Durango and 2.89 times that of Coahuila.

The middle part of Table 9 shows the Economic HH, under two different angles; the Y3 (EEA) and the Y4 (PEA). The FSS shows that producing one US dollar of gain (if the indicator is positive) or loss (if it is negative) implied using 1.82 m^3 per USD of loss in Durango, 0.47 m^3 per USD of loss in Durango and

2.55 m^3 per USD profit in Chihuahua. Once again, Chihuahua had an efficient way of using water in economic terms, since its use allowed it to obtain a profit in relation to the Red Delicious apple in Durango and Chihuahua, where the use of water was not efficient in economic terms, since its use generated losses. In terms of ADP, it is observed that when using the same volume of water, one m^3 , the three states had different economic productivity of water, since while Chihuahua had an ADP of USD 0.39 profit per m^3 , Durango and Coahuila incurred losses of USD 0.55 and USD 2.14 respectively. At the state level, producing one dollar of loss implied using 30.73 m of water, and using one m^3 of water generated USD 0.03 of loss.

At the municipal level, no notable differences are observed for Durango and Coahuila, but not for Chihuahua, since Cuauhtémoc with 3.24 m^3 per USD of profit used more water than the state average (2.55 m^3 per USD of profit), also with USD 0.31 profit per m^3 of water, it had a lower EAP than the state average of USD 0.39. See Table 9.

Finally, the lower part of Table 9, the Social Water Efficiency, abbreviated as SWE (Y5), and the Social Water Productivity, abbreviated as WSP (Y6), shows for SWE, that to produce one job, 65 829, 13 091 and 18 680 m^3 of water had to be invested, i.e., that was the water cost per job generated, and visualized as its inverse, the PES index, measured as jobs per hm^3 of water used in production, shows that when using the same volume of water, one hm^3 , they had different levels of social productivity, since they generated 15.19, 76.39 and 53.53 jobs per hm^3 respectively, finding now that Coahuila was the state with the best level of water efficiency and productivity in social terms, which is explained because that state was where more work days were used, more work hours invested per hectare, that is, technologically it is more backward than Chihuahua, where human labor is substituted to a greater extent by production technification.

5.2 Discussion

The schematic summary of the PFA, PEA and PSA figures and the EFA, EEA and ESA cited in the literature review part of this work and that were analyzed in Table 5, where their indices appear in *absolute* terms, are the lower part of Table 10 where they appear but already in *relative* terms, that is, they are contrasted against the results obtained indicated in Table 9, whose figures appear in the upper part of Table 10, so that the figures in Table 10 come from dividing the figures in Table 5 by those of the Chihuahua Red Delicious apple, so that any resulting indicator, if it is greater than the unit, will indicate that the crop being compared against the Chihuahua Red Delicious apple has a higher productivity (physical, economic or social, as the case may be) of water, for example, Table 5 shows that the Chihuahua pomegranate analyzed by Rios, Hernández and Azpilcueta (2022, *op. cit.*) had a higher water productivity index than the Chihuahua Red Delicious apple (2022, *op. cit.*). *Cit.*) had an ADP index of USD 0.903 profit/m^3 , so that when contrasted against the USD 0.39 profit of the Red Delicious apple from Chihuahua, it was found that: 0.903/0.39=2.30[56] , that is, the pomegranate produces 130% (2.30 - 1.00=1.30=1.30=130%) *more* profit per m^3 of water than the Red Delicious apple from Chihuahua. Those in blue are precisely those crops with indexes lower than unity, i.e., those crops in which the Red Delicious apple is more productive when u s i n g water.

---

[56] The Excel used to obtain each of the tables, as in this case Table 10, closes decimal figures. For this reason, when the indicated quotient of 0.903/0.39 is made, it actually yields 2.315 and not 2.30 as shown in Table 10, thus 0.903 and 0.39 are figures that Excel closed

Table 10: Discussion: Contrast between water productivity indices of Chihuahua Red Delicious apple and other crops. Blue indicates that Red Delicious apple is more productive when using water than Chihuahua Red Delicious versus other crops.
 which compares

LocationSourcePFA	(kg/ m³) EAP (USD/PSA (Jobs/hm³)		

Apple Red Delicious Chihuahua (tér This work 2.557 $ 0.39 53.53

Apple Red Delicious **Chihuahua = 1.0**This work	1.	001.	001	. 00

Contrasting Red Delicious apple indices for the three main producing states/other crops

Apple MT (all the	Chihuahua	Gamboa, Rios and Ros, 2022	0.892	1.58	0.64
Walnut MT	Chihuahua	Gamboa et al, 2022	0.059	0.33	0.17
Grenada	Chihuahua	Rios, Hernandez and Azpilcueta, 2022	0.630	2.30	0.76
Peanut	Chihuahua	Amaya, Rios and Chávez, 2021	0.286	0.48	0.35
Blackberry	Michoacán	Rios, Hernandez and Chavez, 2021	0.563	5.54	2.08
Strawberry	Michoacán	Rios, Hernandez and Chavez, 2021	3.195	9.17	1.45
Bovine milk	Chihuahua	Rios, Rios y Rios, 2019		0.01	
Cotton	Chihuahua	Rios, Pizarro and Galván, 2021	0.140	0.15	0.22
Watermelon	La Laguna, Mexico	Armendariz, Rios & Rodriguez, 2021	5.627	1.43	1.08

Based on the above, it should be noted that in Table 10 the physical productivity indexes lower than that of the Chihuahua Red Delicious apple are shown in blue: MT apple (MT=produced under conditions of Medium Use of Technology) from Chihuahua (with 0.892), pecan nut from Chihuahua (with 0.059), pomegranate from Chihuahua (with 0.630), and peanut (with 0.286).

[3]In a similar situation we observe the blackberry crop from Chihuahua (with 0.563), and cotton from Chihuahua (with 0.140); all these crops had a lower AFP than the Red Delicious apple, with indexes that indicate that the walnut had the lowest relative AFP, because with its index of 0.059 already mentioned, it suggests that when it uses the same amount of water as the Red Delicious apple it produces 2.557 kg of biomass, it will generate only 5.9% of that amount of biomass.557 kg of biomass, it will generate only 5.9% of that amount of biomass, and the one that came closest to the PFA of the Red Delicious apple from Chihuahua was the apple (all varieties) produced in MT with 0.892, which suggests that this crop produces 89.2% of the biomass produced by the Red Delicious apple when both crops use the same volume of water.

The crops that enjoyed a higher AFP than the Chihuahua Red Delicious apple were strawberry from Michoacán and watermelon produced in La Laguna, whose indices equal to 3.195 and 5.627 suggest that when those crops use the same volume of water used by the Chihuahua Red Delicious apple in production, they will yield 219.5% (=3.195 - 1=2.295) and 462.7% (=5.627 - 1 = 4.627).
more biomass. See Table 10.

The crops that had a higher ADP than the Red Delicious apple from Chihuahua were MT apple from Chihuahua (with 1.58), pomegranate from Chihuahua (with 2.30), blackberry and strawberry from Michoacán (with 5.54 and 9.19) and watermelon from La Laguna (with 1.43), observing that in the case of the MT apple from Chihuahua 58% more profit is produced when using the same volumetric unit of water used by the Red Delicious apple from Chihuahua, while the case with the highest relative ADP was the strawberry with 9.17, that is, when using the same volume of water used by the Red Delicious apple it will produce 8.17 times *more* gain or what is the same, it will produce a mass of gain equal to 9.17 times the size of the gain of the Red Delicious apple from Chihuahua. See Table 10

As for the relative PES, Table 10 shows in blue those crops with a lower PES than the Red Delicious apple from Chihuahua. Thus, when these crops use the same volume of water used by the comparison parameter apple analyzed in this work, they will generate less than the 53.33 jobs generated by the parameter, as was the case of MT apple and MT walnut with 0.64 and 0.17, as well as pomegranate with 0.76, peanut with 0.35, and cotton with 0.22. The one with the highest withdrawal from the PES of the Red Delicious apple from Chihuahua was MT walnut, since its index, equal to 0.17, suggests that when the same volume of water is used in the PES of the Red Delicious apple from Chihuahua, it will generate less than the 53.33 jobs generated by the parameter.17, suggests that when the same volume of water is used in production as that used by the Red Delicious apple, only 17% of the employment that this volume of water would produce in the Red Delicious apple from Chihuahua will be produced, and the crop that came closest without surpassing the Red Delicious apple was the pomegranate with 0.76, which suggests that when both crops use the same volume of water, the pomegranate will produce a volume of employment equal to only 76% of that produced by the Red Delicious apple produced with this volume of water. Only blackberry, strawberry from Michoacán and watermelon from La Laguna had a higher PES than the Red Delicious apple from Chihuahua, since their indices were 2.08, 1.45 and 1.08 respectively, suggesting that when using the same volume of water, blackberry, strawberry and watermelon will produce 108% (=2.08 - 1=1.08), 45% (=1.45 - 1=0.45) and 8% (=1.08 - 1 =0.08).

respectively **more employment** than the Red Delicious apple from Chihuahua.

VI CONCLUSIONS AND RECOMMENDATIONS

6.1 Conclusions

The objectives of determining ***the amount of water*** used in commercial-scale production per kg of Red Delicious apple, as well as the amount of water used per USD of profit and the amount of water associated with the creation of a job in the production of Red Delicious apples in the states of Durango, Chihuahua and Coahuila in northern Mexico were met, the objective of determining ***the amount of water*** used in production on a commercial scale per kg of physical product was also met, as well as the amount of water used per USD of profit and the amount of water associated with the creation of a job in the production of Red Delicious apples in the three states of northern Mexico in the agricultural year 2022, therefore the objective of determining the water footprint in its three facets was met: In physical terms, in economic terms and in social terms.

Based on the methodology used, based on the use of mathematical equations that were fed with commercial scale data from SIAP-SADER 2023 (closing of the 2022 agricultural cycle), and the results obtained with such methodological procedure, ***the first hypothesis is not rejected***, since only the Red Delicious apple from the state of Chihuahua, Mexico, with 389 L kg^{-1} , had a lower physical HH than the world average of 700 L kg-1 reported by Mekonnen and Hoekstra (2010 *Op. Cit.*). *Cit.*), Durango and Coahuila had a physical HH well above the world average: 6,346 and 1,111 L kg^{-1} respectively.

Based on the methodology used, based on the use of mathematical equations fed with commercial scale data from SIAP-SADER 2023 (closing of the 2022 agricultural cycle), and the results obtained with such methodological procedure, ***the second hypothesis is not rejected***, since only the Red Delicious apple from the state of Chihuahua, Mexico, with 2.55 m^3 per USD of gain and USD 0.39 gain per m^3 had a higher water efficiency and productivity than the state average of economic water footprint with the corresponding indicators of 39.73 m^3 per USD of loss and USD 0.33 loss per m of water used in production. Red Delicious apples produced in Durango and Coahuila had lower water efficiency and economic water productivity than the state average, in fact, they incurred a loss per m^3 .

Based on the methodology used, based on the use of mathematical equations fed with commercial scale data from SIAP-SADER 2023 (closing of the 2022 agricultural cycle), and the results obtained with such methodological procedure, ***the third hypothesis is*** rejecte*d*, since the state average ESA of Red Delicious apple was 22,570 m^3 Employment^{-1} and the PSA was equal to 44.31 Employment m^{-3} , and both the ESA ***and*** PSA of Chihuahua (with 13,091 m) and the PSA of Chihuahua (with 13,091 m) are equal to 44.31 Employment m , and the ESA and PSA of Chihuahua (with 13,091 m) are equal to 44.31 Employment m .^{3}
/Employment and 76.39 Jobs/hm^3) and Coahuila (with 18,680 m^3 /Employment and
53.53 Jobs/hm^3) were below the state average, but Durango, with 65,829 m^3 /Employment and 15.19 Jobs/hm^3 , had lower efficiency and social productivity of water used in production.

6.2 Recommendations

The government of the state of Chihuahua has a window of opportunity in the world trade of the Red Delicious apple, since the water used in the production of this fruit has high rates of efficiency and physical, economic and social productivity in the water used in its production, characterized, for example, as mentioned above, with a physical HH equivalent to only

55% of the world average physical HH (700 liters/kg), It is therefore important to seek the necessary measures to facilitate exports of this apple to European countries, as they are the most conscious in the trade of products with a low water footprint, thus having a low impact on the environment.

This is reinforced by the fact that the Chihuahua Red Delicious apple has a very favorable social water footprint, as it generates a high number of jobs per volume of water used in its production. It is considered that this aspect would greatly help in the penetration of consumer market niches characterized by the acquisition of products with a high social benefit in their production.

VII LITERATURE CITED

Amaya, L. V. M., Rios, F. J. L., Chávez, R. J. A. 2021. Efectywnosc wykorzystania wody w produkcji orzeszków ziemnuch (Arachis hipogaea L.) w Chihuahua. WYDAWNICTWO NASZA WIEDZA. ISBN 9786203316445 Riga, Latvia.

2000 Agro Agro Industrial Magazine, 2021. Available at: https://www.2000agro.com.mx/agroindustria/crecimiento-de-2- digitos-en-produccion-de-manzanas/

Armandáriz, E. S., Rios, F. J. L. Rodríguez, S. Y. 2021. Rentabilité et utilisation de l'eau dans la culture de la pastèque (*Citrillus lanatus L.*) goutte à goutte à La Laguna, Mexique. EDITIONS NOTRE SAVOR. ISBN 9786203473803. Riga, Latvia

Botanical-oline, 2023. apples, fruit. Available at: https://www.botanical-online.com/botanica/manzano-characteristics#Characteristics_of_apples

BUPA (British United Provident Association). 2023. All the benefits of eating apples. Available at

https://www.bupasalud.com/agentes/novedades/actualizaciones-2023

Cerdán, C. and Suárez, A. I.. 2020. Biodiversity Centers of Origin. Available at: . https://www.uv.mx/personal/asuarez/files/2020/05/Semana-5- Sesion-1-Centros-de-origen-2020.pdf.

Chapagain, A. K. and Hoekstra A.Y. 2004. Water Footprints of Nations. Volume 1: Main Report. Research Report Series No. 16.

DATA MEXICO, GOVERNMENT OF MEXICO, 2023. Available at:

https://www.economia.gob.mx/datamexico/es/profile/product/ap ples-
pears-and-quinces-
fresh#:~:text=In%202021%20at%20world%20level,United%2
0(US%24949M)

Currency converter: available in:

https://www.xe.com/es/currencyconverter/convert/?Amount=1&From=
USD&To=MXN

David, R. 1959. Principles of Political Economy and Taxation. Fondo de
Cultura Económica. First Spanish Edition. Mexico, DF.

El País, 2023. Cetes: the financial instrument that attracts savers in
times of high inflation in Mexico. Last consulted on February 9, 2023.
Available at: https://elpais.com/mexico/2023-02-11/cetes-el-
instrumento- financiero-que-atrae-ahorradores-en-tiempos-de-tiempos-
de-alta-inflacion- en-mexico.html.

.*FAOSTAT, 2021.* ———— International Year of   Fruits and
Vegetables.

Available at: https://www.fao.org/3/cb2395es/cb2395es.pdf

FAOSTAT,   2023.   FAO   Statistics   Division.   Available   at.
https://www.fao.org/3/cb2395es/cb2395es.pdf: accessed   November   8,
2023.

FIRA, 2023, Agrocostos, available in:

https://www.fira.gob.mx/Nd/Agrocostos.jsp

**Gamboa, N. M., Rios, F. J. L., Rios, A. B. E., Rios, A. B., Rios, F. J.
L.. E.. 2022.** Efficiency and productivity of water used in pecan nut (*Carya
illinoensis*) versus apple (*Malus domestica*) production in Chihuahua,
Mexico. Editorial Académica Española. ISBN 978603887532. Berlin,

Germany.

Haro, G. A. Moreau, B. M. Del C. 2023. Apple. In: PULEVA; Bienestarparadisfrutardelavida . Disponible . en. https://www.lechepuleva.es/aprende-a-cuidarte/tu-alimentacion- de-la-a-z/m/manzana#:~:text=The%20apple%20brings%20vitamins%20to%20us%20for%20strengthening%20hair%20and%20u%20C3%B1as

Hoekstra, A.Y. 2003. Virtual Water Trade: Proceedings of the International Expert Meeting on Virtual Water Trade. Delft. The Netherlands. 12 and 13 December 2002. Value of Water Research Report Series No. 12. UNESCO-IHE. Delft. The Netherlands. www.waterfootprint.org/Reports/Report12.pdf.

INIFAP-CENID-RASPA. (2006). Irrigation Program. [Date of consultation:

December 1, 2023]. Available at:

https://cenidraspa.org/serg/serg_v1.php

Marx, C. 1982. Capital. Volume I. FCE, Mexico DF, 1982

Mekonnen, M.M. and Hoekstra, A. Y. 2010. The green, blue and grey water footprint of crops and derived crop products. Hydrology and Earth System Aciences, 15(5): 1577-1600.

Mekonnen, M. M. & Hoekstra, A, Y. 2012. A global Assesment of the wáter footprint of farm animal products. Ecosystems (2012) 15:401-415. DOI:10.1007/s10021-011-9517-8

Rios, F., J. L., Torres M., M. Castro F., Rafael, Torres M., M.A. Ruiz, T. J. 2015. Determination of the blue water footprint in forage crops of DR017 Comarca Lagunera, Mexico. Rev. FCA UNCUYO, 2018. 47(1): 101-122, ISSN print 0370-4661. ISSN (online) 1853-8665, pp.93-107. Mendoza, Argentina.

Rios, F., J. L., Torres M. M. and Torres M., M. A. (2016a). Agricultural water productivity in pecan walnut in northern Mexico. Cases: Comarca Lagunera and Delicias, Chihuahua. ISBN978-3- 639-80166-8. Editorial Académica Española. Saarbrucken, Germany.

Rios, F., J. L., Pizarro, Q., A., Galván, C. R. V. 2021. Wassersparende landwirtschaftliche Muster mit Hilfe des Wasser-Fußabdrucks fall von DR005 Delicias, Chihuahua. Verlag Unser Wissen ISBN9786204117300. Berlin, Germany

Ríos, F., J. L., Rios, A., B. E., Cantú, B., J. E., Rios, A., H. E., Armendáriz, E., S., Chávez, R. J. A., Navarrete M., C. & Castro, F., R. 2018. Analysis of physical, economic and social water efficiency in asparagus (*Asparagus officinalis L.*) and grapes (*Vitis* vinifera) table of DR-037 Altar-Pitiquito-Caborca, Sonora, Mexico 2018. *Revista de la Facultad de Ciencias Agrarias. Universidad Nacional de Cuyo*, 50(2). ISSN print 0370-4661, ISSN (online) 1853-8665. Mendoza, Argentina.

Rios, F. J. L., Hernández, I. G. and Azpilcueta R., M.. 2021. Economia agricolo.ambientale dell'acqua utilizzata sulla produzione commerciale di melograno (*Punica granatum L.*) EDIZIONI SAPIENZA. ISBN 9786205168288. Berlino, Germania

Rios, F., J. L., Rios A., Becky E, Rios A., Hebrián E. 2019. Physical and economic water footprints of milk. The case of bovine milk from Delicias, Chihuahua, Mexico. Editorial Académica Española. ISBN 978-620-0-02518-0. Berlin, Germany

Rios, F., J. L., Hernández, I. G. and Chávez, R., J. A. 2021. Economic productivity of water in Blackberry (*Rubus spp. L.*) and strawberry (*Fragaria vesca L.*): The case of Irrigation District 061 in Zamora. OUR KNOWLEDGE PUBLISHING.

ISBN9786203189537. Berlin, Germany

SIAP-SADER (2023). Agricultural closure 2022. Available at: http://infosiap.siap.gob.mx/aagricola_siap_gb/icultivo/

Wikipedia. Available at: https://es.wikipedia.org/wiki/Manzana

Zarza, L., F. 2023. Difference between blue footprint, green footprint and gray footprint. Available at: https://www.iagua.es/respuestas/diferencia-blue-footprint-green-footprint-and-grey-footprint.

yes I want morebooks!

Buy your books fast and straightforward online - at one of world's fastest growing online book stores! Environmentally sound due to Print-on-Demand technologies.

Buy your books online at
www.morebooks.shop

Kaufen Sie Ihre Bücher schnell und unkompliziert online – auf einer der am schnellsten wachsenden Buchhandelsplattformen weltweit! Dank Print-On-Demand umwelt- und ressourcenschonend produziert.

Bücher schneller online kaufen
www.morebooks.shop

Printed by Books on Demand GmbH, Norderstedt / Germany